# Springer Topics in Signal Processing

Volume 12

More information about this series at http://www.springer.com/series/8109

Jacob Benesty · Jingdong Chen
Israel Cohen

# Design of Circular Differential Microphone Arrays

Springer

Jacob Benesty
INRS-EMT
University of Quebec
Montreal, QC
Canada

Israel Cohen
Department of Electrical Engineering
Technion, Israel Institute of Technology
Haifa
Israel

Jingdong Chen
Northwestern Polytechnical University
Xi'an, Shaanxi
China

ISSN 1866-2609 ISSN 1866-2617 (electronic)
Springer Topics in Signal Processing
ISBN 978-3-319-14841-0 ISBN 978-3-319-14842-7 (eBook)
DOI 10.1007/978-3-319-14842-7

Library of Congress Control Number: 2015930029

Springer Cham Heidelberg New York Dordrecht London


Printed on acid-free paper

Springer International Publishing AG Switzerland is part of Springer Science+Business Media (www.springer.com)

# Abstract

Recently, we proposed a completely novel and efficient way to design differential beamforming algorithms with linear microphone arrays. Thanks to this very flexible approach, any order of differential arrays can be designed. Moreover, they can be made robust against white noise amplification, which is the main inconvenience in these types of arrays. The other well-known problem with linear arrays is that electronic steering is not really feasible.

In this book, we extend all these fundamental ideas to circular microphone arrays and show that we can design small and compact differential arrays of any order, which can electronically steer in many different directions, with a good control of the white noise amplification problem, high directional gains, and frequency-independent responses. We give many practical examples, proving that differential beamforming with circular microphone arrays is likely one of the best possible choices for speech enhancement (i.e., noise reduction and dereverberation). Almost, all the material presented in this text is new and will be of great interest to engineers, students, and researchers working in microphone arrays and their applications to all types of telecommunications, security, and surveillance.

# Contents

# Chapter 1
# Introduction

## 1.1 Introduction

In general, a microphone array refers to a sound acquisition system that uses multiple microphones to sample the sound field with spatial diversity. These microphones are arranged into a particular geometry in which each sensor's position relative to a reference point is known to the subsequent processors. This kind of arrays, when combined with proper signal processing algorithms that can take advantage of the spatial, temporal, and frequency information embedded in the observations, can be used to solve many important room acoustic problems such as source localization/tracking, noise reduction/speech enhancement, source separation, dereverberation, spatial sound recording, etc. Consequently, the design of such microphone arrays and the associated processing algorithms has attracted a significant amount of research and engineering interest over several decades.

Some of the fundamental principles of microphone array processing can be traced back to the 1930's when the directional microphones were invented [1], [2]. The formal terminology of "microphone array" however, appears in the 1960's. Early works in this area were strongly influenced by the sensor array theory developed in the field of radar and sonar processing. It was soon found that the problems encountered in microphone arrays are very different from those in radar and sonar. The causes that lead to the difference are multiple and the major ones are as follows. 1) Microphone arrays deal with speech and audio signals, which are broadband in nature and have a frequency range from a few Hz to 20 kHz. It is extremely challenging, if not impossible, to develop algorithms that can perform uniformly over such a large frequency range. 2) Speech and audio signals are highly nonstationary; as a result, the estimation of their statistics is very difficult. 3) The environments where microphone arrays are supposed to work are generally reverberant and it is not uncommon to have thousands of reflections. It is well known that reverberation can have a very negative impact on the performance of microphone array processing

 
J. Benesty et al., *Design of Circular Differential Microphone Arrays*,
Springer Topics in Signal Processing 12, DOI 10.1007/978-3-319-14842-7_1

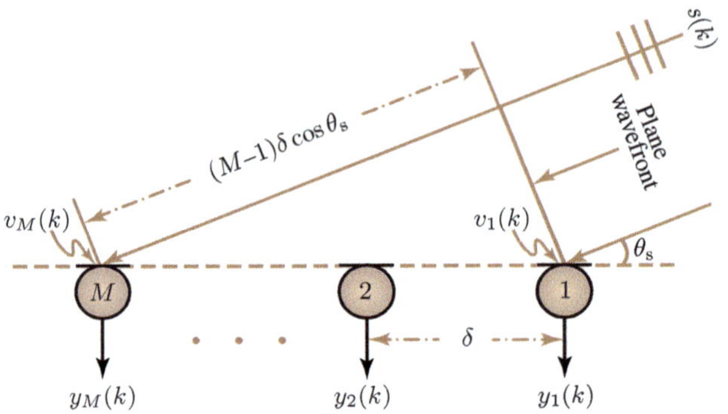


**Fig. 1.1** Illustration of a uniform linear microphone array.

algorithms. 4) The human ear has an extremely wide dynamic range (as much as 120 dB for normal hearing) and is very sensitive to weak distortion caused by processing algorithms. 5) The number of sensors in microphone arrays is usually not that large. As a consequence, how to achieve much better performance with microphone arrays than with simple single-channel systems is a challenging task. Because of these reasons, the many algorithms that work fairly well for applications of sensor arrays fail to perform in microphone arrays. Therefore, lots of efforts have been made in the last couple of decades to make microphone arrays work.

## 1.2 General Principles of Microphone Arrays

In a microphone array, each sensor's output consists of not only the signal of interest, but also noise, interference, and reverberation. With proper processing of all the sensors' outputs, the signal of interest can be extracted while unwanted noise, interference, and late reflections can be suppressed.

To illustrate how a microphone array deals with additive noise, let us consider a simple example with a uniform linear array (ULA) of $M$ microphones as shown in Fig. 1.1. Assume that there is a single desired source in the farfield such that its spherical wavefront appears planar at the array. In the anechoic case and if we neglect the propagation attenuation, the signals received at the $M$ microphones, at the discrete-time index $k$, can be written as

$$\begin{aligned} y_m(k) &= x_m(k) + v_m(k) \\ &= s(k - t - \tau_m) + v_m(k) \\ &= x(k - \tau_m) + v_m(k), \ m = 1, 2, \ldots, M, \end{aligned} \tag{1.1}$$

where $t$ is the propagation time from the source $s(k)$ to microphone 1 (reference point), $\tau_m$ is the relative time delay between the $m$th microphone and the first one, and $x_m(k) = s(k - t - \tau_m) = x(k - \tau_m)$ and $v_m(k)$ are, respectively, the desired and noise signals at the $m$th microphone. The noise signal, $v_m(k)$, is assumed to be uncorrelated with $s(k)$. Moreover, all signals are assumed to be zero mean. With a ULA and a farfield source, the delay $\tau_m$ can be expressed into the following form according to the geometry shown in Fig. 1.1:

$$\tau_m = (m - 1)\frac{\delta \cos \theta_{\mathrm{s}}}{c}, \ m = 1, 2, \ldots, M, \tag{1.2}$$

where $\delta$ is the spacing between two neighboring microphones, $c$ is the sound velocity in the air, and $\theta_{\mathrm{s}}$ is the source incidence angle. For simplicity of exposition, let us work in the frequency domain. Then, the signal model in (1.1) is written as

$$\begin{aligned} Y_m(\omega) &= X_m(\omega) + V_m(\omega) \\ &= S(\omega)e^{-\jmath\omega(t+\tau_m)} + V_m(\omega) \\ &= X(\omega)e^{-\jmath\omega\tau_m} + V_m(\omega), \ m = 1, 2, \ldots, M, \end{aligned} \tag{1.3}$$

where $\omega = 2\pi f$ is the angular frequency, $f$ denotes frequency, $\jmath$ represents the imaginary unit with $\jmath^2 = -1$, $Y_m(\omega)$, $X_m(\omega)$, $V_m(\omega)$, and $S(\omega)$ are the Fourier transform of $y_m(k)$, $x_m(k)$, $v_m(k)$, and $s(k)$ respectively, and $X(\omega) = S(\omega)e^{-\jmath\omega t} = X_1(\omega)$.

Now, let us process the $M$ observation signals, $Y_m(\omega)$, $m = 1, 2, \ldots, M$, in order to extract the desired signal and reduce the effect of the additive noise via linear beamforming, i.e.,

$$\begin{aligned} Z(\omega) &= \sum_{m=1}^{M} H_m^*(\omega)Y_m(\omega) \\ &= \mathbf{h}^H(\omega)\mathbf{y}(\omega), \end{aligned} \tag{1.4}$$

where $Z(\omega)$ is an estimate of the desired signal, the superscripts $^*$ and $^H$ are the complex conjugation and conjugate-transpose operator, respectively,

$$\mathbf{h}(\omega) = \begin{bmatrix} H_1(\omega) \ H_2(\omega) \cdots \ H_M(\omega) \end{bmatrix}^T \tag{1.5}$$

is the so-called beamforming filter, the superscript $^T$ is the transpose operator, and

$$\mathbf{y}(\omega) = \left[ Y_1(\omega)\ Y_2(\omega) \cdots\ Y_M(\omega) \right]^T. \tag{1.6}$$

The simplest beamformer is the so-called delay-and-sum (DS) in which the beamforming filter coefficients are chosen as $H_m(\omega) = \frac{1}{M} e^{-\jmath \omega \tau_m}$, $m = 1, 2, \ldots, M$. In other words, the principle of the DS beamformer is that we compensate $y_m(k)$ with a delay $\tau_m$ to align all the $M$ microphone signals and then average the results together. The DS beamformer's output is then

$$\begin{aligned} Z(\omega) &= \frac{1}{M} \sum_{m=1}^{M} Y_m(\omega) e^{\jmath \omega \tau_m} \\ &= X(\omega) + \frac{1}{M} \sum_{m=1}^{M} V_m(\omega) e^{\jmath \omega \tau_m}. \end{aligned} \tag{1.7}$$

To check whether the beamformer's output is less noisy than its input, let us compare the input and output signal-to-noise ratios (SNRs). The input SNR of the DS beamformer, according to the signal model given in (1.3), is defined as the SNR at the first (reference) microphone, i.e.,

$$\mathrm{iSNR}(\omega) = \frac{\phi_{X_1}(\omega)}{\phi_{V_1}(\omega)} = \frac{\phi_X(\omega)}{\phi_{V_1}(\omega)}, \tag{1.8}$$

where $\phi_{X_1}(\omega) = E\left[|X_1(\omega)|^2\right]$, $\phi_{V_1}(\omega) = E\left[|V_1(\omega)|^2\right]$, and $\phi_X(\omega) = E\left[|X(\omega)|^2\right]$, with $E[\cdot]$ denoting mathematical expectation. The output SNR of the beamformer, from (1.7), is written as

$$\mathrm{oSNR}\left[\mathbf{h}(\omega)\right] = \frac{\phi_X(\omega)}{\frac{1}{M^2} E\left[\left|\sum_{m=1}^{M} V_m(\omega) e^{\jmath \omega \tau_m}\right|^2\right]}. \tag{1.9}$$

Now, if all the noise signals, $V_m(\omega)$, $m = 1, 2, \ldots, M$, are mutually uncorrelated and have the same variance, it can be checked that $\mathrm{oSNR}\left[\mathbf{h}(\omega)\right] = M \times \mathrm{iSNR}(\omega)$. In this case, a simple DS beamformer can improve the SNR by a factor of $M$ or $10 \log_{10} M$ in dB. This gain in SNR is independent of the signal frequency.

One interesting way to evaluate the array performance is via the beampattern. For the signals given in (1.3), if we neglect the noise terms and choose $H_m(\omega) = \frac{1}{M} e^{-\jmath \omega (m-1)\delta \cos\theta / c}$, the DS beamformer's output is then

$$Z(\omega) = \frac{1}{M} \sum_{m=1}^{M} X(\omega) e^{-\jmath \omega (m-1)\delta(\cos\theta_s - \cos\theta)/c}. \tag{1.10}$$

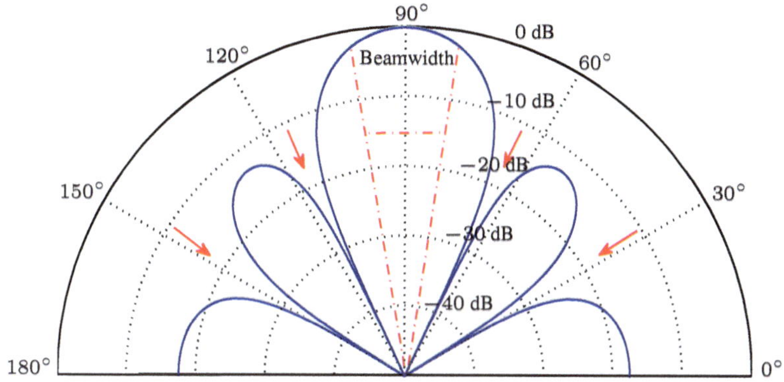


**Fig. 1.2** Beampattern of the DS beamformer as a function of $\theta$ with a ULA. $M = 10$, $\theta_s = 90°$, $\delta = 4$ cm, and $f = 2$ kHz.

The beampattern, which is defined as the magnitude of the transfer function between the beamformer's output and the input signal, is then

$$\mathcal{B}(\omega, \theta, \theta_s) = \left| \frac{Z(\omega)}{X(\omega)} \right| = \left| \frac{\sin\left[M\omega\delta(\cos\theta_s - \cos\theta)/(2c)\right]}{M \sin\left[\omega\delta(\cos\theta_s - \cos\theta)/(2c)\right]} \right| . \tag{1.11}$$

Figure 1.2 plots the beampattern [eq. (1.11)] for $M = 10$, $\delta = 4$ cm, $\theta_s = 90°$, and $f = 2$ kHz. The resulting beampattern consists of a total of 5 beams in the range between 0° and 180°. The one with the highest amplitude (0 dB) is called the main lobe and all the others are called side lobes. One important parameter regarding the main lobe is the beamwidth (sometimes also called main lobe width), which is defined as the angle region between the −3-dB points on either side of the main lobe. The height of the side lobes represents the gain pattern for noise and competing sources present along the directions other than the mainlobe direction. In array and beamforming design, one can hope to make the side lobes as low as possible so that signals coming from directions other than the main lobe direction will be attenuated as much as possible. In addition, there are a number of directions where $\mathcal{B}(\omega, \theta, \theta_s) = 0$. These are called nulls of the beampattern, which can be used to optimize beamforming filters as will be discussed throughout the text.

For the DS beamformer, the beampattern is a function of the number of sensors, the interelement spacing, the source incidence angle, and the signal frequency. Regarding the spacing, the array beamwidth decreases as the spacing $\delta$ increases. So, if we want a sharper beam, we can simply increase $\delta$. However, when the spacing is increased to a value larger than $\lambda/2 = c/(2f)$, where $\lambda$ is the wavelength of the signal, spatial aliasing arises. To visualize this problem, we plot the beampattern of the DS beamformer with a ULA for

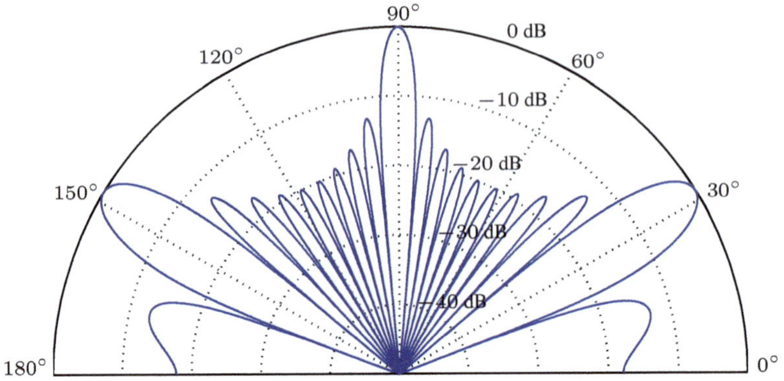


**Fig. 1.3** Beampattern of the DS beamformer as a function of $\theta$ with a ULA. $M = 10$, $\theta_s = 90°$, $\delta = 20$ cm, and $f = 2$ kHz.

$M = 10$, $\theta_s = 90°$, and $f = 2$ kHz. But this time, the interelement spacing is 20 cm. The corresponding beampattern is shown in Fig. 1.3. We can see three large beams that have a maximum amplitude of 1 (0 dB). The other two (than the main lobe) are called grating lobes. Signals propagating from directions at which grating lobes occur would be indistinguishable from signals propagating from the main lobe direction. This ambiguity is often referred to as spatial aliasing. In order to avoid this problem, the interelement spacing has to satisfy $\delta \leq \lambda/2 = c/(2f)$.

The above simple DS beamformer, though illustrated with a ULA, can be used with any geometry such as the circular, planar, and spherical ones. However, when the DS beamformer processes broadband speech signals, it suffers from a number of problems and drawbacks. 1) Its beampattern is frequency dependent and the beamwidth is inversely proportional to the frequency as shown in Fig. 1.4. Therefore, this beamformer is not effective in dealing with low-frequency noise and interference. 2) Noise is not uniformly attenuated over its entire spectrum, resulting in some disturbing artifacts in the array output [5]. 3) If the incident angle of the speech source is different from the array's look direction, the speech signal will be low-pass filtered, leading to speech distortion. To overcome these drawbacks, the so-called broadband beamforming techniques have been investigated.

One way to obtain a broadband beamformer is to use harmonically nested subarrays where every subarray is designed for operating at a single frequency [6], [7]. Similar beamwidth can be obtained across the frequency range of interest by controlling the spacing and the number of sensors in every subarray. But such a solution requires a large array aperture with many microphones, even though different subarrays may share sensors in the array. Another way

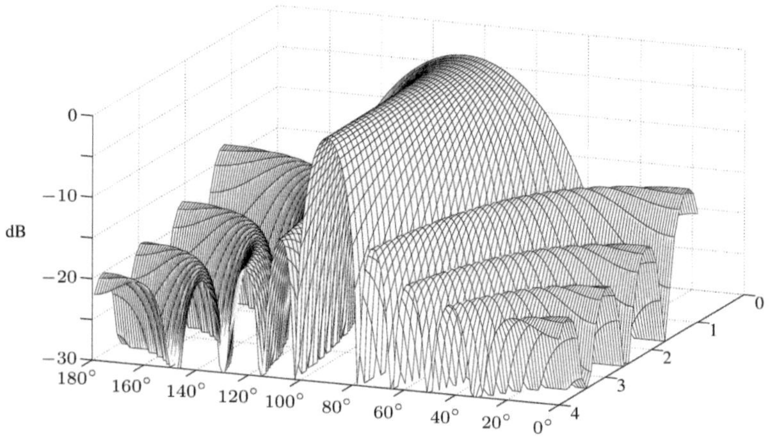


**Fig. 1.4** 3D beampattern of the DS beamformer as a function of $f$ and $\theta$ with a ULA. $M = 10$, $\theta_s = 90°$, and $\delta = 4$ cm.

to design a broadband beamformer is based on narrowband decomposition as illustrated in Fig. 1.5(a). A narrowband beamformer is then designed in each subband with a constraint applied to control the beamwidth so that all the beamformers from different subbands have the same beamwidth. Though it can make similar beamwidth across a wide range of frequencies, this way of broadband beamforming sacrifices the array performance in high frequencies. The subband structure of broadband beamforming can be equivalently transformed into its time-domain counterpart shown in Fig. 1.5(b), where an FIR filter is applied to each sensor output, and the filtered sensor signals are then added up together to form a single output. This is widely known as the filter-and-sum beamformer first developed by Frost in [8]. The core problem in this technique is to determine the coefficients of those FIR filters. These coefficients can be determined by using some statistics models of the noise field. The resulting beamformer is called fixed beamformer. Alternatively, they can be determined based on some *a priori* knowledge of either the acoustic environment or the array data, leading to a class of adaptive beamformers such as the the minimum variance distortionless response (MVDR) and linearly constrained minimum variance (LCMV) filters [8], [9, 10, 3]. Generally, adaptive beamformers can be more efficient than the fixed ones in suppressing directional noise and competing sources; but they may suffer from signal self cancelation, which deserves careful attention.

While those narrowband and filter-and-sum structures are often referred to as "broadband" beamformers, they are not truly "broadband" in the sense that their beampatterns still vary significantly with frequency. A more proper

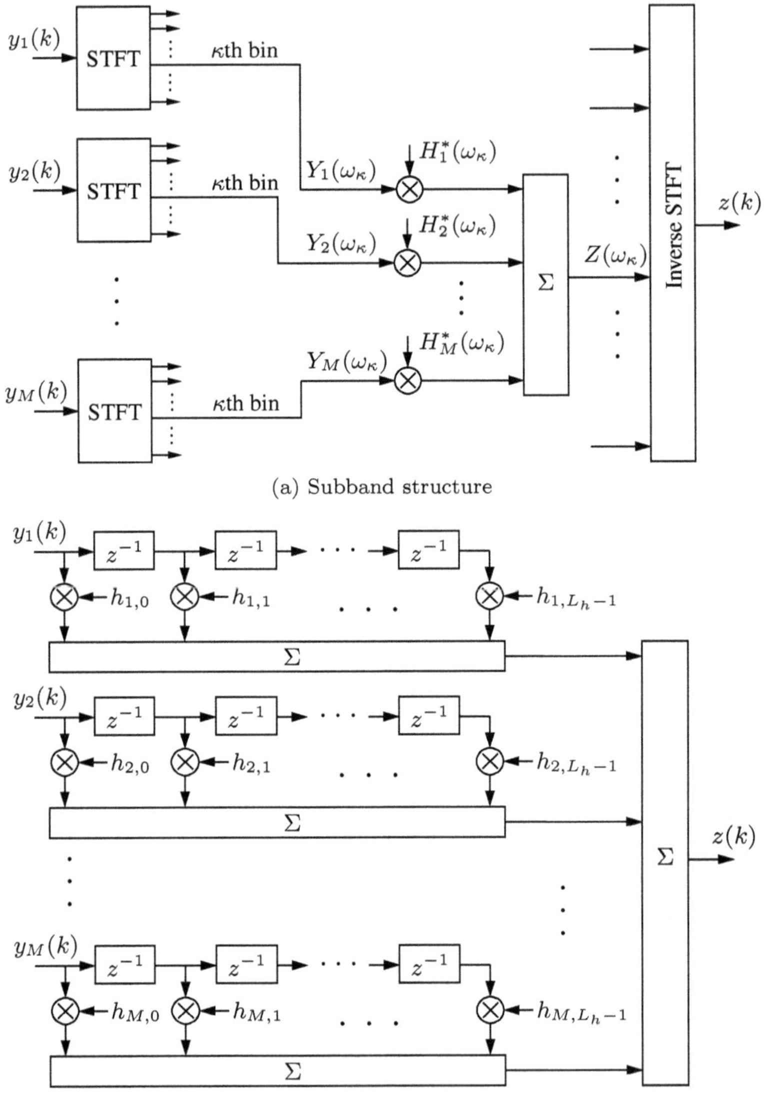


(a) Subband structure



(b) Filter-and-sum structure

**Fig. 1.5** Broadband beamforming Structures.

way to design broadband beamformers is perhaps with differential microphone arrays (DMAs), which can have frequency-invariant beampatterns. DMAs refer to the arrays that are responsive to the spatial derivatives of the acoustic pressure field. The basic idea of such arrays started from the directional ribbon microphones [1], [2] and was then extended to forming the widely used directional electret (capacitor) microphones [11], [12]. Directional microphones, in their early form, have a prominent limitation, i.e., once made, their directional response is fixed. If, for a given application, we find that the microphone we bought does not produce the expected performance, the only option is to try a different type. There is no flexibility in adapting the directional pattern to fit the application needs. To circumvent this limitation, the modern concept of DMAs was developed, in which a number of pressure microphones are arranged into a particular geometry and digital signal processing techniques are then used to process the microphones' outputs to obtain the desired directional response [13]–[28]. Figure 1.6 illustrates how first-, second-, and third-order DMAs are constructed with a linear geometry. Basically, a general $n$th-order DMA has a response proportional to a linear combination of signals derived from spatial derivatives from order 0 to (including) order $n$. Note that an inherent assumption in the construction of a DMA is that the microphones are placed close enough so that the true acoustic pressure differentials can be approximated by finite differences between microphone sensors' outputs or, in other words, the spacing between neighboring microphones should be small in DMAs. The small spacing brings some good advantages to the array. For example, it makes the beampatten frequency-invariant, so the array can have a uniform response at a wide range of frequencies. It also makes the array small in size, which can be integrated into small communications devices. However, the small spacing may also bring some negative effect such as white noise amplification, particularly in low frequencies.

Although DMAs have many attractive properties, the conventional way of designing them as shown in Fig. 1.6 suffers from a number of problems. Not only it lacks flexibility in forming different patterns, it is also difficult to deal with the problem of white noise amplification. Recently, we developed a new approach to the design and implementation of uniform linear DMAs [4, 29, 30]. The basic structure is similar to the one in Fig. 1.5(a). The DMA beamforming is achieved in three steps. First, the signals received by the DMA are partitioned into small overlapping frames and every frame is then transformed into the short-time Fourier transform (STFT) domain. In every STFT subband, a differential beamformer is designed by solving a linear system formed from some fundamental constraints on the null information of the desired beampatterns. This designed DMA beamformer is applied to the multichannel STFT coefficients to reduce noise and estimate the desired signal in the corresponding subband. Finally, the time-domain signal estimate is constructed using either the overlap-add or overlap-save technique with the inverse STFT. The major advantages that differentiate this new DMA design approach from the traditional one are the following. 1) It is much more flexible to design different patterns by using only the null information. 2) It is easy to analyze the array performance. 3) A minimum-norm approach is formed that

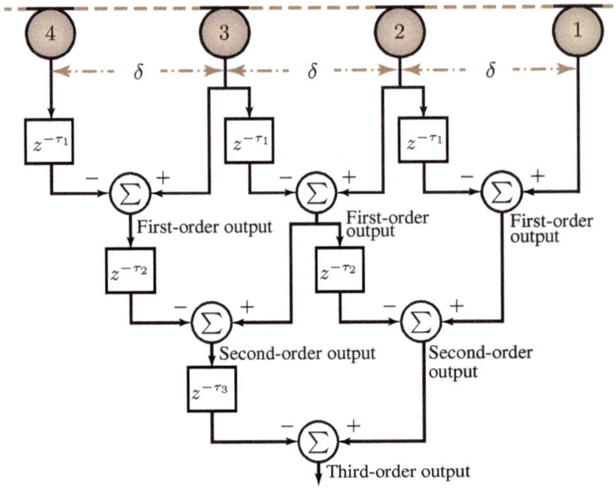


**Fig. 1.6** Conventional structure of first-, second-, and third-order DMAs.

can maximize the white noise gain with a given number of sensors, so the white noise amplification problem can be addressed by increasing the number of microphones. In this book, we extend all the fundamental ideas developed in [4, 29, 30] for linear differential microphones arrays to circular differential microphone arrays. This extension is, actually, not that straightforward.

## 1.3 From Linear Differential Arrays to Circular Differential Arrays

The geometry of microphone arrays plays an important role in the formulation, solution, and performance of the processing algorithms. The selection of the geometry, however, depends heavily on the application requirements. For example, in devices like smartphones, tablet PCs, and smart televisions, the linear geometry is preferable as this type of arrays can be easily integrated into these devices. But linear arrays may not have the same response at different directions as shown in [31]. In applications like teleconferencing and 3D sound recording where the signal of interest many come from any direction, it is necessary for the microphone array to have similar, if not the same, response from one direction to another. In this case, circular arrays are often used. Another important advantage of using circular arrays is that the processing problem can be greatly simplified due to the symmetry property. This book addresses the beamforming problem with circular microphone arrays. We discuss the design of small and compact circular differential microphone arrays of any order with frequency-invariant patterns, high directional gains, and good control of white noise amplification. Moreover, the arrays can be electronically steered to many different directions without affecting the performance.

## 1.4 Organization of the Book

The material in this book is organized into eight chapters, including this one. In the next seven chapters, we attempt to cover the most basic concepts and fundamental techniques used in the design and implementation of the different orders of circular differential arrays and the associated beamforming algorithms. All is explained from a signal processing perspective. The material discussed in these chapters is as follows.

Chapter 2 presents the general formulation of beamforming and differential beamforming in the context of circular microphone arrays. It introduces several basic concepts that are important and useful in the design and evaluation of beamformers, particularly differential beamformers, including the steering vector for a plane wave with the conventional anechoic farfield model, the beampattern and the DMA beampattern of different orders, the gain in SNR, the white noise gain, the directivity factor, etc.

Chapter 3 is dedicated to the design of first-order circular differential microphone arrays. It discusses how to form the dipole, cardioid, subcardioid, hypercardioid, and supercardioid patterns with three sensors. It also illustrates the impact of interelement spacing on the beampattern, white noise gain, and directivity factor.

The design of second-order circular differential microphone arrays is investigated in Chapter 4. This chapter covers the design in two cases: four and five sensors. Again, examples are provided to show how to design the most well-known directivity patterns including the dipole, cardioid, hypercardioid, and supercardioid.

Chapter 5 starts with the design of third-order circular differential microphone arrays with either six or seven sensors and then generalizes the principle to the design of any order. Examples are presented to illustrate the design of third-order patterns with three distinct nulls.

Chapter 6 is concerned with the design of superdirective beamformers. It covers both linear and circular arrays. It also discusses the inherent relationship between superdirective and differential beamformers. They both fit into the category of beamforming with supergains.

One of the most critical issue with DMAs is white noise amplification, particularly in low frequencies. In [4], a minimum-norm approach was developed, which can take advantage of more microphones (the number of microphones is larger than the DMA order plus 1) to maximize the white noise gain. In Chapter 7, the same approach is extended to differential circular arrays.

The pattern of a DMA can be obtained from the general definition of the beampattern by approximating the exponential function with the Jacobi-Anger expansion. This approximation can be used to design circular differential arrays, which is discussed in Chapter 8.

## References

1. H. F. Olson, "A uni-directional ribbon microphone," *J. Acoust. Soc. Am.*, vol. 3, pp. 315, 1932.
2. H. F. Olson, "Gradient microphones," *J. Acoust. Soc. Am.*, vol. 17, pp. 192–198, Jan. 1946.
3. J. Benesty, J. Chen, and Y. Huang, *Microphone Array Signal Processing.* Berlin, Germany: Springer-Verlag, 2008.
4. J. Benesty and J. Chen, *Study and Design of Differential Microphone Arrays.* Berlin, Germany: Springer-Verlag, 2012.
5. D. B. Ward, R. C. Williamson, and R. A. Kennedy, "Broadband microphone arrays for speech acquisition," *Acoust. Australia*, vol. 26, pp. 17-20, Apr. 1998.
6. J. L. Flanagan, J. D. Johnson, R. Zahn, and G. W. Elko, "Computer-steered microphone arrays for sound transduction in large rooms," *J. Acoust. Soc. Amer.*, vol. 75, pp. 1508–1518, Nov. 1985.
7. J. L. Flanagan, D. A. Berkley, G. W. Elko, J. E. West, and M. M. Sondhi, "Autodirective microphone systems," *Acustica*, vol. 73, pp. 58–71, Feb. 1991.
8. O. L. Frost, III, "An algorithm for linearly constrained adaptive array processing," *Proc. IEEE*, vol. 60, pp. 926–935, Aug. 1972.
9. B. Widrow, P. Mantey, L. Griffiths, and B. Goode, "Adaptive antenna systems," *Proc. IEEE*, vol. 55, pp. 2143–2159, Dec. 1967.
10. J. Capon, "High resolution frequency-wavenumber spectrum analysis," *Proc. IEEE*, vol. 57, pp. 1408–1418, Aug. 1969.
11. G. M. Sessler and J. E. West, "Directional transducers," *IEEE Trans. Audio Electroacuostic.*, vol. 19, pp. 19–23, Mar. 1971.
12. P. Graven and M. Gerzon, "Coincident microphone simlation covering three dimensional space and yielding various directional outputs," U.S. Patent 4,042,779, Aug. 1977.
13. G. Elko, "Steerable and variable first order differential microphone array," U.S. Patent 6,041,127, Apr. 1997.
14. G. W. Elko and A.-T. N. Pong, "A steerable and variable first-order differential micropone array," in *Proc. IEEE ICASSP*, 1997.
15. G. W. Elko, "Superdirectional microphone arrays," in *Acoustic Signal Processing for Telecommunication*, S. L. Gay and J. Benesty, Eds. Boston, MA: Kluwer Academic Publishers, 2000, Chapter 10, pp. 181–237.
16. G. W. Elko and J. Meyer, "Microphone arrays," in *Springer Handbook on Speech Processing and Speech Communication*, J. Benesty, M. M. Sondhi, and Y. Huang, Eds. Berlin, Germany: Springer-Verlag, 2007.
17. M. Williamsen, "Design and measurement of a dipole microphone," *AudioXpress*, pp. 1–8, vol. 7, 2009.
18. H. Teutsch and G. W. Elko, "First- and second-order adaptive differential microphone arrays," in *Proc. IWAENC*, 2001
19. J. Pekonen, "Microphone techniques for spatial sound," in *Proc. the 2008 Acoustics Seminar on Spatial Sound Modeling*, 2008.

20. T. D. Abhayapala and A. Gupta, "Higher order differential-integral microphone arrays," *J. Acoust. Soc. Am.*, vol. 127, pp. EL227-233, May 2010.
21. M. Buck and M. Rößler, "First order differential microphone arrays for automotive applications," in *Proc. IWAENC*, 2001.
22. J. Bitzer, K.-D. Kammeyer, and K. U. Simmer, "An alternative implementation of the superdirective beamformer," in *Proc. IEEE WASPAA*, 1999.
23. M. Buck, "Aspects of first-order differential microphone arrays in the presence of sensor imperfections," *European Trans. Telecommunications*, vol. 13, pp. 115–122, 2002.
24. E. D. Sena, H. Hac1habiboğlu, and Z. Cvetković, "On the design and implementation of higher-order differential microphones," *IEEE Trans. Audio, Speech, Lang. Process.*, vol. 20, pp. 162–174, Jan. 2012.
25. V. Hamacher, J. Chalupper, J. Eggers, E. Fischer, U. Kornagel, H. Puder, and U. Rass, "Signal processing in high-end hearing aids: state of the art, challenges, and future trends," *EURASIP J. Applied Signal Process.*, vol. 2005, pp. 2915–2929, 2005.
26. R. Derkx, "Optimal azimuthal steering of a first-order supperdirectional microphone response," in *Proc. IWAENC*, 2008.
27. R. N. Miles, Y. Liu, Q. Su, and E. Cui, "A silicon directional microphone with second-order directivity," in *Proc. inter. Congress Acoust.*, 2007.
28. M. Kolundžija, C. Faller, and M. Vetterli, "Spatio-temporal gradient analysis of differential microphone arrays," in *Proc. Audio Engineering Society Convention*, 2009.
29. J. Chen and J. Benesty, "A general approach to the design and implementation of linear differential microphone arrays," in *Proc. APSIPA Annual Summit and Conference*, 2013.
30. J. Chen, J. Benesty, and C. Pan "On the design and implementation of linear differential microphone arrays," *J. Acoust. Soc. Am.*, vol. 136, pp. 3097–3113, Dec. 2014.
31. C. Pan, J. Chen, and J. Benesty, "Performance study of the MVDR beamformer as a function of the source incident angle," *IEEE/ACM Trans. Audio, Speech, Lang. Process.*, vol. 22, pp. 67–79, Jan. 2014.

# Chapter 2
# Problem Formulation

In this chapter, we explain some important aspects of beamforming and differential arrays with a focus on the circular geometry. The problem of a DMA design is formulated while we progress in defining some useful concepts. We start with the definition of the steering vector for a plane wave with the conventional anechoic farfield model, which has an interesting structure. We give the general definition of the beampattern as well as its expression for differential arrays. We explain a fundamental property, which is the basis for the design of circular differential microphone arrays. We then derive the gain in signal-to-noise ratio (SNR), which is very useful in the evaluation of DMAs under different types of noise.

## 2.1 Signal Model

We consider a source signal (plane wave), in the farfield, that propagates in an anechoic acoustic environment at the speed of sound, i.e., $c = 340$ m/s, and impinges on a uniform circular array (UCA), of radius $r$, consisting of $M$ omnidirectional microphones. The direction of the source signal to the array is parameterized by the azimuth angle $\theta$. We assume that the center of the UCA coincides with the origin of the Cartesian coordinate system, azimuth angles are measured anti-clockwise from the $x$ axis, i.e., at $\theta = 0$, and sensor 1 of the array is placed on the $x$ axis, i.e., at $\theta = 0$ (see Fig. 2.1). When operating in the farfield, the time delay between microphone $m$ and the center of the array is given by [1]

$$\tau_m = \frac{r}{c} \cos(\theta - \psi_m), \ m = 1, 2, \ldots, M, \tag{2.1}$$

where

 
J. Benesty et al., *Design of Circular Differential Microphone Arrays*,
Springer Topics in Signal Processing 12, DOI 10.1007/978-3-319-14842-7_2

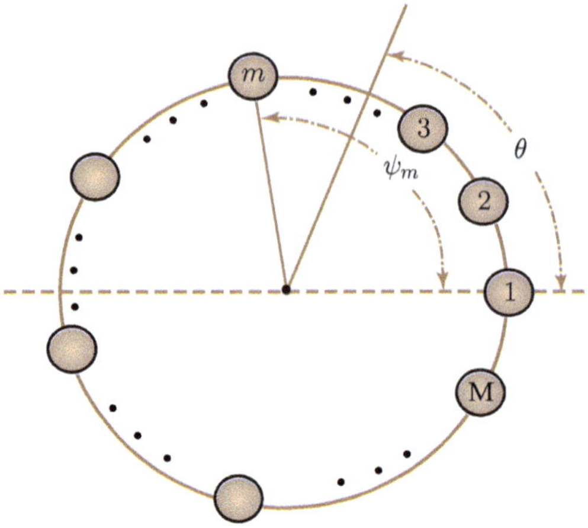


**Fig. 2.1** Illustration of a uniform circular microphone array in the Cartesian coordinate system.

$$\psi_m = \frac{2\pi(m-1)}{M} \tag{2.2}$$

is the angular position of the $m$th array element. In this scenario, the steering vector of length $M$ is

$$\begin{aligned} \mathbf{d}(\omega, \theta) &= \left[ e^{\jmath\omega\tau_1} \; \cdots \; e^{\jmath\omega\tau_M} \right]^T \\ &= \left[ e^{\jmath\omega r c^{-1}\cos(\theta-\psi_1)} \; \cdots \; e^{\jmath\omega r c^{-1}\cos(\theta-\psi_M)} \right]^T, \end{aligned} \tag{2.3}$$

where the superscript $^T$ is the transpose operator, $\jmath = \sqrt{-1}$ is the imaginary unit, $\omega = 2\pi f$ is the angular frequency, and $f > 0$ is the temporal frequency. The acoustic wavelength is $\lambda = c/f$. For a UCA, the interelement spacing is

$$\begin{aligned} \delta &= 2r\sin\left(\frac{\pi}{M}\right) \\ &\approx \frac{2\pi r}{M}. \end{aligned} \tag{2.4}$$

In order to avoid spatial aliasing [2], which has the negative effect of creating grating lobes (i.e., copies of the main lobe, which usually points toward the desired signal), it is necessary that the interelement spacing is less than $\lambda/2$, i.e.,

$$M > \frac{4\pi r}{\lambda} = \frac{4\pi r f}{c}. \tag{2.5}$$

The condition (2.5) easily holds for small values of $r$ (or $\delta$) and in low frequencies but not in high frequencies.

Substituting the approximation of (2.4) into (2.3), the steering vector can be rewritten as

$$\mathbf{d}(f,\theta) \approx \left[ e^{\jmath f \delta c^{-1} \cos(\theta - \psi_1)} \ \ldots \ e^{\jmath M f \delta c^{-1} \cos(\theta - \psi_M)} \right]^T, \quad (2.6)$$

which is interesting to compare to the steering vector of a uniform linear array (ULA):

$$\mathbf{d}_{\mathrm{L}}(\omega,\theta) = \left[ 1 \ e^{-\jmath \omega \delta c^{-1} \cos\theta} \ \ldots \ e^{-\jmath (M-1) \omega \delta c^{-1} \cos\theta} \right]^T. \quad (2.7)$$

There are two fundamental differences between these two steering vectors. The first one is that (2.6) depends on $\cos(\theta - \psi_m)$ while (2.7) depends on $\cos\theta$. The other difference is that $\mathbf{d}(f,\theta) \neq \mathbf{d}(f,-\theta)$ while $\mathbf{d}_{\mathrm{L}}(\omega,\theta) = \mathbf{d}_{\mathrm{L}}(\omega,-\theta)$. The fact that $\mathbf{d}(\omega,\theta)$ has no apparent symmetry makes the design of circular differential microphone arrays (CDMAs) very different from the design of linear differential microphone arrays (LDMAs). However, we still have an interesting symmetry for the entries of $\mathbf{d}(\omega,\theta)$.

*Property 2.1.* For the angle $\theta = 0$, let

$$\mathbf{d}(\omega,0) = \left[ D_1(\omega,0) \ D_2(\omega,0) \ \cdots \ D_M(\omega,0) \right]^T. \quad (2.8)$$

We have

$$D_{m+1}(\omega,0) = D_{M-m+1}(\omega,0), \ m = 1,2,\ldots,M-1. \quad (2.9)$$

*Proof.* It is easy to check that the following relations hold

$$\cos(\psi_{m+1}) = \cos(\psi_{M-m+1}), \ m = 1,2,\ldots,M-1. \quad (2.10)$$

Therefore, (2.9) is true.

For some particular values of $\theta$, the steering vector of a UCA has another interesting property.

*Property 2.2.* For the angular positions, $\psi_m$, $m = 1,2,\ldots,M$, of the $M$ sensors of a UCA, we have

$$\mathbf{d}(\omega,\psi_m) = \mathbf{P}_m \mathbf{d}(\omega,0), \quad (2.11)$$

where $\mathbf{P}_m$ is an $M \times M$ permutation matrix whose form is

$$\mathbf{P}_m = \mathbf{P}_2^{m-1}, \ m = 1,2,\ldots,M, \quad (2.12)$$

where $\mathbf{P}_1 = \mathbf{I}_M$, with $\mathbf{I}_M$ being the $M \times M$ identity matrix, and

$$\mathbf{P}_2 = \begin{bmatrix} 0\,0 & \cdots & 0\,0\,1 \\ 1\,0 & \cdots & 0\,0\,0 \\ 0\,1 & \cdots & 0\,0\,0 \\ \vdots\,\vdots & \ddots & \vdots\,\vdots\,\vdots \\ 0\,0 & \cdots & 1\,0\,0 \\ 0\,0 & \cdots & 0\,1\,0 \end{bmatrix}. \tag{2.13}$$

*Proof.* This property can be shown by using (2.9) and the relations:

$$\cos\left(\psi_i - \psi_j\right) = \cos\left(\psi_{|i-j|+1}\right),\ i, j = 1, 2, \ldots, M. \tag{2.14}$$

Since $\mathbf{P}_m$ is a permutation matrix, we have

$$\mathbf{P}_m^T \mathbf{P}_m = \mathbf{P}_m \mathbf{P}_m^T = \mathbf{I}_M. \tag{2.15}$$

As a result,

$$\mathbf{d}\left(\omega, 0\right) = \mathbf{P}_m^T \mathbf{d}\left(\omega, \psi_m\right). \tag{2.16}$$

To simplify the presentation of the equations in the rest of this text, we will also use the variable:

$$\varpi = \frac{\omega r}{c}. \tag{2.17}$$

Let us denote by $\theta_{\mathrm{s}}$ the steering angle of the array. We consider designing fixed directional beamformers[1], like in DMAs, where the main lobe points at $\theta = \theta_{\mathrm{s}}$ and the desired signal propagates from the same angle. We recall that for LDMAs, the optimal position is at $\theta = 0$ or $\pi$ (endfire direction) and electronic steering (in the sense that the main lobe can be oriented to any possible direction without affecting the shape of the beampattern) is not really feasible. We will see that with CDMAs, we have much more flexibilities.

As pointed out in [3], there is a fundamental difference between differential arrays and filter-and-sum beamformers. In the latter category, the filters are optimized in such a way that the microphone signals are aligned in order to steer the main lobe in the direction of the desired signal, whereas in the former category the gains are optimized to steer a number of nulls in some specific directions.

The focus of this work is on the design, with small circular apertures, of beamformers whose beampatterns are very close to the ones obtained with "ideal" DMAs but in the direction $\theta_{\mathrm{s}}$. For that, a complex weight, $H_m^*\left(\omega, \theta_{\mathrm{s}}\right),\ m = 1, 2, \ldots, M$, is applied at the output of each microphone, where the superscript $^*$ denotes complex conjugation. The weighted outputs are then summed together to form the beamformer output. Putting all the

[1] The terms beamformer, beamforming, and beampattern may not be adequate in the context of DMAs but we will still use them for convenience.

gains together in a vector of length $M$, we get

$$\mathbf{h}(\omega, \theta_s) = \left[ H_1(\omega, \theta_s) \; H_2(\omega, \theta_s) \; \cdots \; H_M(\omega, \theta_s) \right]^T . \tag{2.18}$$

Then, the objective is to design such a filter for any directivity pattern of any order when the array geometry is circular. The approach taken here is based on the fundamental observation that for all beampatterns of interest, some constraints must be fulfilled at all frequencies given that the number of microphones is equal to $M$. In other words, we select a certain number of fundamental constraints from a well-defined beampattern of a DMA to design $\mathbf{h}(\omega, \theta_s)$. The case $M = 2$ has no interest in this investigation since it is equivalent to a ULA, which has been extensively studied in the literature, and corresponds to a first-order DMA [4], [5].

In the next two sections, we discuss some fundamental measures. We are only interested in narrowband measures. The broadband measures can be easily deduced from their respective narrowband counterparts [6].

## 2.2 Beampattern

Each beamformer has a pattern of directional sensitivity, i.e., it has different sensitivities from sounds arriving from different directions. The beampattern or directivity pattern describes the sensitivity of the beamformer to a plane wave (source signal) impinging on the UCA from the direction $\theta$. Mathematically, it is defined as

$$\begin{aligned} \mathcal{B}\left[\mathbf{h}(\omega, \theta_s), \theta\right] &= \mathbf{h}^H(\omega, \theta_s)\, \mathbf{d}(\omega, \theta) \qquad (2.19) \\ &= \sum_{m=1}^{M} H_m^*(\omega, \theta_s)\, e^{\jmath \varpi \cos(\theta - \psi_m)}, \end{aligned}$$

where the superscript $^H$ is the conjugate-transpose operator.

The frequency-independent beampattern of an $N$th-order DMA is well known. Its definition is usually given for a steering angle of 0 [4]. For any steering angle, $\theta_s$, this beampattern is defined as

$$\mathcal{B}_N(\theta - \theta_s) = \sum_{n=0}^{N} a_{N,n} \cos^n(\theta - \theta_s), \tag{2.20}$$

where $a_{N,n}$, $n = 0, 1, \ldots, N$, are real coefficients. The different values of these coefficients determine the different directivity patterns of the $N$th-order DMA. In the direction of the desired signal, i.e., for $\theta = \theta_s$, the beampattern must be equal to 1, i.e., $\mathcal{B}_N(0) = 1$. Therefore, we have

$$\sum_{n=0}^{N} a_{N,n} = 1. \tag{2.21}$$

As a result, we always choose the first coefficient as

$$a_{N,0} = 1 - \sum_{n=1}^{N} a_{N,n}. \tag{2.22}$$

The most interesting patterns have at least one null in some direction. We also have the important property:

$$\mathcal{B}_N(\theta - \theta_s) = \mathcal{B}_N(-\theta + \theta_s). \tag{2.23}$$

The directivity pattern of any DMA is an even function. Therefore, on a polar plot[2], $\mathcal{B}_N(\theta - \theta_s)$ is symmetric about the axis $\theta_s; \theta_s + \pi$ and any DMA design can be restricted to this range. It follows from (2.20) that an $N$th-order DMA has at most $N$ (distinct) nulls in this range.

The directivity factor (see also Section 2.3) of an $N$th-order DMA, defined as the ratio between the directivity pattern at the direction $\theta = \theta_s$ and the averaged directivity pattern over the whole space, is[3] [4], [7], [8]

$$\begin{aligned} \mathcal{G}_N &= \frac{\mathcal{B}_N^2(0)}{\dfrac{1}{\pi}\displaystyle\int_{\theta_s}^{\theta_s+\pi} \mathcal{B}_N^2(\theta - \theta_s)\, d\theta} \\ &= \frac{\pi}{\displaystyle\int_{\theta_s}^{\theta_s+\pi} \left[\sum_{n=0}^{N} a_{N,n} \cos^n(\theta - \theta_s)\right]^2 d\theta} \end{aligned} \tag{2.24}$$

and what we call the directivity index is

$$\mathcal{D}_N = 10 \log_{10} \mathcal{G}_N. \tag{2.25}$$

We find that the first-order, second-order, and third-order directivity factors are

[2] Polar patterns are a very convenient way to describe the directional sensitivity of the DMAs.

[3] This situation corresponds to the cylindrically isotropic noise field.

$$\mathcal{G}_1 = \frac{1}{a_{1,0}^2 + \frac{1}{2}a_{1,1}^2}, \tag{2.26}$$

$$\mathcal{G}_2 = \frac{1}{a_{2,0}^2 + \frac{1}{2}a_{2,1}^2 + \frac{3}{8}a_{2,2}^2 + a_{2,0}a_{2,2}}, \tag{2.27}$$

$$\mathcal{G}_3 = \frac{1}{a_{3,0}^2 + \frac{1}{2}a_{3,1}^2 + \frac{3}{8}a_{3,2}^2 + a_{3,0}a_{3,2} + \frac{5}{16}a_{3,3}^2 + \frac{3}{4}a_{3,1}a_{3,3}}. \tag{2.28}$$

The hypercardioid is the pattern obtained from the maximization of the directivity factor[4].

The front-to-back ratio is defined as the ratio of the power of the output of the array to signals propagating from the front-half plane (rotated by $\theta_s$) to the output power for signals arriving from the rear-half plane (rotated by $\theta_s$) [9]. This ratio, for the cylindrically isotropic noise field, is mathematically defined as [4], [9]

$$\mathcal{F}_N = \frac{\int_{\theta_s}^{\theta_s+\pi/2} \mathcal{B}_N^2 (\theta - \theta_s)\, d\theta}{\int_{\theta_s+\pi/2}^{\pi} \mathcal{B}_N^2 (\theta - \theta_s)\, d\theta}. \tag{2.29}$$

The supercardioid is the pattern obtained from the maximization of the front-to-back ratio[5] [9].

First-order directivity patterns have the form:

$$\mathcal{B}_1 (\theta - \theta_s) = (1 - a_{1,1}) + a_{1,1} \cos (\theta - \theta_s) \tag{2.30}$$

and the most important ones are as follows.

- Dipole: $a_{1,1} = 1$, null at $\cos (\theta - \theta_s) = 0$, and $\mathcal{D}_1 = 3$ dB.
- Cardioid: $a_{1,1} = \frac{1}{2}$, null at $\cos (\theta - \theta_s) = -1$, and $\mathcal{D}_1 = 4.3$ dB.
- Hypercardioid: $a_{1,1} = \frac{2}{3}$, null at $\cos (\theta - \theta_s) = -1/2$, and $\mathcal{D}_1 = 4.8$ dB.
- Supercardioid: $a_{1,1} = 2 - \sqrt{2}$, null at $\cos (\theta - \theta_s) = (1 - \sqrt{2})/(2 - \sqrt{2})$, and $\mathcal{D}_1 = 4.6$ dB.

Figure 2.2 shows these different polar patterns for $\theta_s = 0$. What is exactly shown are the values of the magnitude squared beampattern in dB, i.e., $10 \log_{10} \mathcal{B}_1^2 (\theta)$.

Second-order beampatterns are described by the equation:

---

[4] Another type of hypercardioid can be obtained by maximizing the directivity factor in the presence of a spherically isotropic noise field. There is not much difference, however, between the two patterns.

[5] Another type of supercardioid can be obtained by maximizing the front-to-back ratio in the presence of a spherically isotropic noise field. There is not much difference, however, between the two patterns.

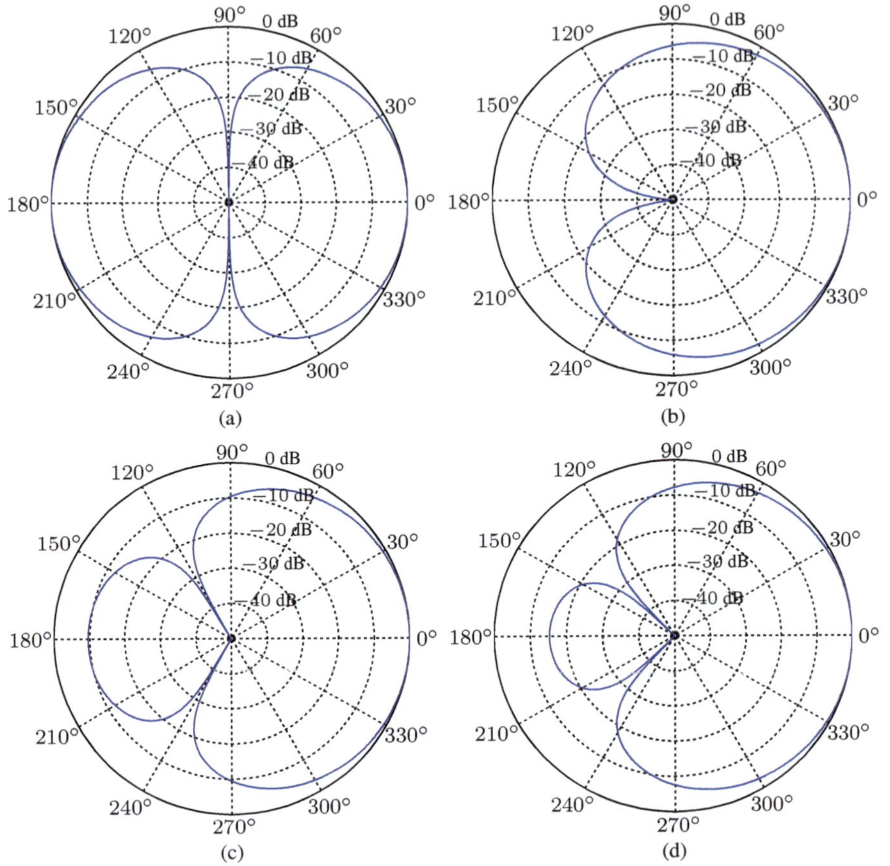


**Fig. 2.2** First-order directivity patterns: (a) dipole, (b) cardioid, (c) hypercardioid, and (d) supercardioid.

$$\mathcal{B}_2\left(\theta-\theta_{\mathrm{s}}\right)=\left(1-a_{2,1}-a_{2,2}\right)+a_{2,1}\cos\left(\theta-\theta_{\mathrm{s}}\right)+a_{2,2}\cos^2\left(\theta-\theta_{\mathrm{s}}\right). \tag{2.31}$$

The second-order dipole has a null at $\cos\left(\theta-\theta_{\mathrm{s}}\right)=0$ and a one (maximum) at $\cos\left(\theta-\theta_{\mathrm{s}}\right)=-1$. Replacing these values in (2.31), we find that $a_{2,1}=0$ and $a_{2,2}=1$. By analogy to the first-order and second-order dipoles, we define the $N$th-order dipole as

$$\mathcal{B}_{\mathrm{D},N}\left(\theta-\theta_{\mathrm{s}}\right)=\cos^N\left(\theta-\theta_{\mathrm{s}}\right), \tag{2.32}$$

implying that $a_{N,N}=1$ and $a_{N,N-1}=a_{N,N-2}=\cdots=a_{N,0}=0$. The $N$th-order dipole has only one (distinct) null (in the range $\theta_{\mathrm{s}};\theta_{\mathrm{s}}+\pi$) at $\theta=\theta_{\mathrm{s}}+\pi/2$. The directivity indices of the second-order and third-order dipoles are, respectively, $\mathcal{D}_2=4.3$ dB and $\mathcal{D}_3=5.1$ dB.

The most well-known second-order cardioid has two nulls; one at $\cos(\theta - \theta_s) = -1$ and the other one at $\cos(\theta - \theta_s) = 0$. From these values, we easily deduce from (2.31) that $a_{2,1} = a_{2,2} = \frac{1}{2}$. By analogy to the first-order and second-order cardioids, we define the $N$th-order cardioid as

$$\mathcal{B}_{\mathrm{C},N}(\theta - \theta_s) = \left[\frac{1}{2} + \frac{1}{2}\cos(\theta - \theta_s)\right]\cos^{N-1}(\theta - \theta_s), \quad (2.33)$$

implying that $a_{N,N} = a_{N,N-1} = \frac{1}{2}$ and $a_{N,N-2} = a_{N,N-3} = \cdots = a_{N,0} = 0$. This $N$th-order cardioid has only two distinct nulls (in the range $\theta_s; \theta_s + \pi$): one at $\theta = \theta_s + \pi/2$ and the other one at $\theta = \theta_s + \pi$. The directivity indices of the second-order and third-order cardioids are, respectively, $\mathcal{D}_2 = 6.6$ dB and $\mathcal{D}_3 = 7.6$ dB.

The $N$th-order hypercardioid and supercardioid are characterized by the fact that they have $N$ distinct nulls in the interval $\theta_s < \theta < \theta_s + \pi$. Hence, their general beampattern is

$$\mathcal{B}_{\mathrm{HS},N}(\theta - \theta_s) = \prod_{n=1}^{N}\left[\varsigma_{N,n} + (1 - \varsigma_{N,n})\cos(\theta - \theta_s)\right]. \quad (2.34)$$

Third-order beampatterns have the form

$$\begin{aligned}\mathcal{B}_3(\theta - \theta_s) = &(1 - a_{3,1} - a_{3,2} - a_{3,3}) + a_{3,1}\cos(\theta - \theta_s) \\ &+ a_{3,2}\cos^2(\theta - \theta_s) + a_{3,3}\cos^3(\theta - \theta_s).\end{aligned} \quad (2.35)$$

We give the values of $a_{N,n}$ and $\mathcal{D}_N$ for some examples of hypercardioid and supercardioid [4], [8]:

- second-order hypercardioid, $a_{2,1} = \frac{2}{5}$, $a_{2,2} = \frac{4}{5}$, $\mathcal{D}_2 = 7$ dB;
- second-order supercardioid, $a_{2,1} \approx 0.484$, $a_{2,2} \approx 0.413$, $\mathcal{D}_2 = 6.3$ dB;
- third-order hypercardioid, $a_{3,1} = -\frac{4}{7}$, $a_{3,2} = \frac{4}{7}$, $a_{3,3} = \frac{8}{7}$, $\mathcal{D}_3 = 8.4$ dB; and
- third-order supercardioid, $a_{3,1} \approx 0.217$, $a_{3,2} \approx 0.475$, $a_{3,3} \approx 0.286$, $\mathcal{D}_3 = 7.2$ dB.

Figures 2.3 and 2.4 depict the different second-order and third-order directivity patterns discussed above for $\theta_s = 0$.

Now, let us get back to the general definition of the UCA beampattern given in (2.19). To make the analysis of a CDMA similar to an LDMA, we assume that $\theta_s = 0$. Then, we can simplify the notation by writing $\mathbf{h}(\omega, \theta_s) = \mathbf{h}(\omega)$. We will discuss the more general case of any steering angle later on. In order to be able to design a CDMA of any order, its beampattern must be an even function like in an LDMA, i.e., we must have

$$\mathcal{B}[\mathbf{h}(\omega), \theta] = \mathcal{B}[\mathbf{h}(\omega), -\theta]. \quad (2.36)$$

In the following, we will use the relations:

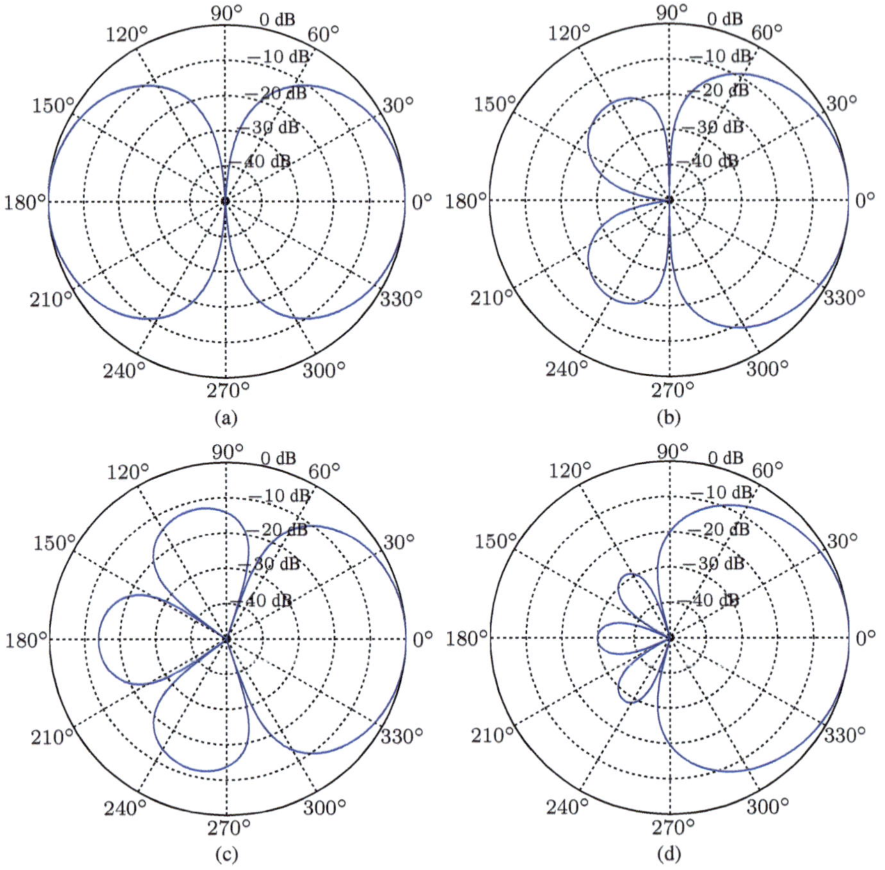


**Fig. 2.3** Second-order directivity patterns: (a) dipole, (b) cardioid, (c) hypercardioid, and (d) supercardioid.

$$\cos(\theta + \psi_m) = \cos(\theta - \psi_{M-m+2}), \; m = 1, 2, \ldots, M. \tag{2.37}$$

For $M = 2$, we have

$$\begin{aligned} \mathcal{B}[\mathbf{h}(\omega), -\theta] &= H_1^*(\omega) e^{\jmath\varpi\cos\theta} + H_2^*(\omega) e^{\jmath\varpi\cos(\theta+\psi_2)} \\ &= H_1^*(\omega) e^{\jmath\varpi\cos\theta} + H_2^*(\omega) e^{\jmath\varpi\cos(\theta-\psi_2)} \\ &= \mathcal{B}[\mathbf{h}(\omega), \theta]. \end{aligned} \tag{2.38}$$

For any positive integer $M$, it is obvious that

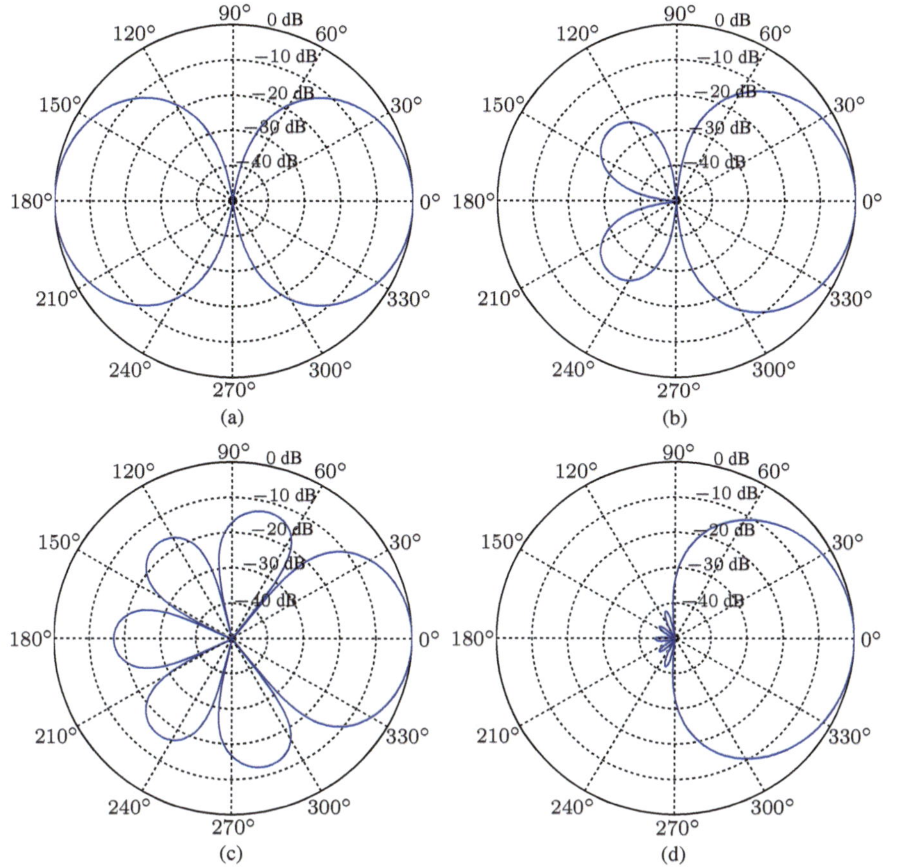


**Fig. 2.4** Third-order directivity patterns: (a) dipole, (b) cardioid, (c) hypercardioid, and (d) supercardioid.

$$\mathcal{B}\left[\mathbf{h}\left(\omega\right), -\theta\right] = \sum_{m=1}^{M} H_m^*\left(\omega\right) e^{\jmath \varpi \cos\left(-\theta - \psi_m\right)} \tag{2.39}$$
$$= \sum_{m=1}^{M} H_m^*\left(\omega\right) e^{\jmath \varpi \cos\left(\theta + \psi_m\right)}.$$

Substituting (2.37) into (2.39), we get

$$\mathcal{B}\left[\mathbf{h}\left(\omega\right), -\theta\right] = \sum_{m=1}^{M} H_m^*\left(\omega\right) e^{\jmath \varpi \cos\left(\theta - \psi_{M-m+2}\right)}. \tag{2.40}$$

Therefore, (2.36) is verified if and only if

$$H_{m+1}(\omega) = H_{M-m+1}(\omega), \ m = 1, 2, \ldots, M-1. \tag{2.41}$$

We observe that in the filter $\mathbf{h}(\omega)$ of length $M$, only its first $\lfloor \frac{M}{2} \rfloor + 1$ coefficients need to be optimized, where $\lfloor x \rfloor$ is the integer part of $x$. This also means that only $\lfloor \frac{M}{2} \rfloor + 1$ independent constraints are possible. As a result, with a UCA of $M$ microphones, we can design any differential array up to the order $\lfloor \frac{M}{2} \rfloor$ at the steering angle $\theta_s = 0$.

It is clear that we can design filters in such a way that the directivity patterns are all identical (up to a rotation) for

$$\theta_s = \psi_m = \frac{2\pi(m-1)}{M}, \ m = 1, 2, \ldots, M, \tag{2.42}$$

because of the perfect alignment of the desired source signal (at $\theta_s = \psi_m$), the position of the $m$th microphone, and the center of the array. Also, from this axis, the beampattern is perfectly symmetric. It is possible to electronically steer the array in other directions (than $\theta_s = \psi_m$), but the shape of the resulting patterns may not resemble anymore to the well-known ones. We can perfectly steer at $\theta_s = \psi_m$ because of Property 2.2.

We give the following fundamental property, which is the basis for the design of CDMAs.

*Property 2.3.* With $M$ microphones and the designed filter, $\mathbf{h}(\omega)$, with the symmetry constraint [eq. (2.41)], we can build a CDMA up to the order $\lfloor \frac{M}{2} \rfloor$, which can perfectly steer in $M$ different directions $\psi_m = \frac{2\pi(m-1)}{M}$, $m = 1, 2, \ldots, M$. We recall that $\psi_m$ is the position of the $m$th microphone. Steering in other directions is also possible but there is no guaranty that the original directivity pattern will not be affected.

In practice, we only need to design the filter for $\theta_s = 0$ with the conditions on the coefficients of $\mathbf{h}(\omega)$ given in (2.41). The filter corresponding to $\theta_s = \psi_m$ is easily obtained by simply permuting the coefficients of the filter designed for $\theta_s = 0$.

## 2.3 Gain in Signal-to-Noise Ratio (SNR)

We recall that the desired signal comes from the angle $\theta_s$. In this case, the $m$th microphone signal is given by

$$Y_m(\omega) = e^{\jmath\varpi\cos(\theta_s - \psi_m)} X(\omega) + V_m(\omega), \ m = 1, 2, \ldots, M, \tag{2.43}$$

where $X(\omega)$ is the desired signal and $V_m(\omega)$ is the additive noise at the $m$th microphone. In a vector form, (2.43) becomes

$$\begin{aligned}\mathbf{y}(\omega) &= \left[Y_1(\omega)\ Y_2(\omega)\ \cdots\ Y_M(\omega)\right]^T \\ &= \mathbf{d}(\omega, \theta_s) X(\omega) + \mathbf{v}(\omega),\end{aligned} \tag{2.44}$$

where $\mathbf{d}(\omega, \theta_s)$ is the steering vector at $\theta = \theta_s$ and the noise signal vector, $\mathbf{v}(\omega)$, is defined similarly to $\mathbf{y}(\omega)$.

The beamformer output is simply

$$\begin{aligned}Z(\omega) &= \sum_{m=1}^{M} H_m^*(\omega, \theta_s) Y_m(\omega) \\ &= \mathbf{h}^H(\omega, \theta_s)\mathbf{y}(\omega) \\ &= \mathbf{h}^H(\omega, \theta_s)\mathbf{d}(\omega, \theta_s) X(\omega) + \mathbf{h}^H(\omega, \theta_s)\mathbf{v}(\omega),\end{aligned} \tag{2.45}$$

where $Z(\omega)$ is an estimate of the desired signal, $X(\omega)$.

If we consider the first microphone as the reference, we can define the input signal-to-noise ratio (SNR) with respect to this reference as

$$\mathrm{iSNR}(\omega) = \frac{\phi_X(\omega)}{\phi_{V_1}(\omega)}, \tag{2.46}$$

where $\phi_X(\omega) = E\left[|X(\omega)|^2\right]$ and $\phi_{V_1}(\omega) = E\left[|V_1(\omega)|^2\right]$ are the variances of $X(\omega)$ and $V_1(\omega)$, respectively, with $E[\cdot]$ denoting mathematical expectation.

The output SNR is defined as

$$\begin{aligned}\mathrm{oSNR}\left[\mathbf{h}(\omega, \theta_s)\right] &= \phi_X(\omega)\frac{\left|\mathbf{h}^H(\omega, \theta_s)\mathbf{d}(\omega, \theta_s)\right|^2}{\mathbf{h}^H(\omega, \theta_s)\mathbf{\Phi}_{\mathbf{v}}(\omega)\mathbf{h}(\omega, \theta_s)} \\ &= \frac{\phi_X(\omega)}{\phi_{V_1}(\omega)} \times \frac{\left|\mathbf{h}^H(\omega, \theta_s)\mathbf{d}(\omega, \theta_s)\right|^2}{\mathbf{h}^H(\omega, \theta_s)\mathbf{\Gamma}_{\mathbf{v}}(\omega)\mathbf{h}(\omega, \theta_s)},\end{aligned} \tag{2.47}$$

where

$$\mathbf{\Phi}_{\mathbf{v}}(\omega) = E\left[\mathbf{v}(\omega)\mathbf{v}^H(\omega)\right] \tag{2.48}$$

and

$$\mathbf{\Gamma}_{\mathbf{v}}(\omega) = \frac{\mathbf{\Phi}_{\mathbf{v}}(\omega)}{\phi_{V_1}(\omega)} \tag{2.49}$$

are the correlation and pseudo-coherence matrices of $\mathbf{v}(\omega)$, respectively.

The definition of the gain in SNR is easily derived from the previous definitions, i.e.,

$$\mathcal{G}\left[\mathbf{h}\left(\omega,\theta_{\rm s}\right)\right] = \frac{\text{oSNR}\left[\mathbf{h}\left(\omega,\theta_{\rm s}\right)\right]}{\text{iSNR}\left(\omega\right)} = \frac{\left|\mathbf{h}^H\left(\omega,\theta_{\rm s}\right)\mathbf{d}\left(\omega,\theta_{\rm s}\right)\right|^2}{\mathbf{h}^H\left(\omega,\theta_{\rm s}\right)\boldsymbol{\Gamma}_{\mathbf{v}}\left(\omega\right)\mathbf{h}\left(\omega,\theta_{\rm s}\right)}. \tag{2.50}$$

Assume that the matrix $\boldsymbol{\Gamma}_{\mathbf{v}}(\omega)$ is nonsingular. In this case, for any two vectors $\mathbf{h}(\omega,\theta_{\rm s})$ and $\mathbf{d}(\omega,\theta_{\rm s})$, we have

$$\left|\mathbf{h}^H\left(\omega,\theta_{\rm s}\right)\mathbf{d}\left(\omega,\theta_{\rm s}\right)\right|^2 \leq \left[\mathbf{h}^H\left(\omega,\theta_{\rm s}\right)\boldsymbol{\Gamma}_{\mathbf{v}}\left(\omega\right)\mathbf{h}\left(\omega,\theta_{\rm s}\right)\right] \times \left[\mathbf{d}^H\left(\omega,\theta_{\rm s}\right)\boldsymbol{\Gamma}_{\mathbf{v}}^{-1}\left(\omega\right)\mathbf{d}\left(\omega,\theta_{\rm s}\right)\right], \tag{2.51}$$

with equality if and only if $\mathbf{h}(\omega,\theta_{\rm s}) \propto \boldsymbol{\Gamma}_{\mathbf{v}}^{-1}(\omega)\,\mathbf{d}(\omega,\theta_{\rm s})$. Using the inequality (2.51) in (2.50), we deduce an upper bound for the gain:

$$\begin{aligned}\mathcal{G}\left[\mathbf{h}\left(\omega,\theta_{\rm s}\right)\right] &\leq \mathbf{d}^H\left(\omega,\theta_{\rm s}\right)\boldsymbol{\Gamma}_{\mathbf{v}}^{-1}\left(\omega\right)\mathbf{d}\left(\omega,\theta_{\rm s}\right)\\ &\leq \text{tr}\left[\boldsymbol{\Gamma}_{\mathbf{v}}^{-1}\left(\omega\right)\right]\text{tr}\left[\mathbf{d}\left(\omega,\theta_{\rm s}\right)\mathbf{d}^H\left(\omega,\theta_{\rm s}\right)\right]\\ &\leq M\text{tr}\left[\boldsymbol{\Gamma}_{\mathbf{v}}^{-1}\left(\omega\right)\right],\end{aligned} \tag{2.52}$$

where tr[$\cdot$] is the trace of a square matrix. We observe how the gain is upper bounded [as long as $\boldsymbol{\Gamma}_{\mathbf{v}}(\omega)$ is nonsingular] and depends on the number of microphones as well as on the nature of the noise.

In our context, the distortionless constraint is desired, i.e.,

$$\mathbf{h}^H\left(\omega,\theta_{\rm s}\right)\mathbf{d}\left(\omega,\theta_{\rm s}\right) = 1. \tag{2.53}$$

As a consequence, it is easy to see that the filter:

$$\mathbf{h}_{\max}\left(\omega,\theta_{\rm s}\right) = \frac{\boldsymbol{\Gamma}_{\mathbf{v}}^{-1}\left(\omega\right)\mathbf{d}\left(\omega,\theta_{\rm s}\right)}{\mathbf{d}^H\left(\omega,\theta_{\rm s}\right)\boldsymbol{\Gamma}_{\mathbf{v}}^{-1}\left(\omega\right)\mathbf{d}\left(\omega,\theta_{\rm s}\right)} \tag{2.54}$$

maximizes the gain, which is given by

$$\mathcal{G}\left[\mathbf{h}_{\max}\left(\omega,\theta_{\rm s}\right)\right] = \mathbf{d}^H\left(\omega,\theta_{\rm s}\right)\boldsymbol{\Gamma}_{\mathbf{v}}^{-1}\left(\omega\right)\mathbf{d}\left(\omega,\theta_{\rm s}\right). \tag{2.55}$$

We are interested in three types of noise.

- The temporally and spatially white noise with the same variance at all microphones[6]. In this case, $\boldsymbol{\Gamma}_{\mathbf{v}}(\omega) = \mathbf{I}_M$. Therefore, the white noise gain is

$$\mathcal{G}_{\text{wn}}\left[\mathbf{h}\left(\omega,\theta_{\rm s}\right)\right] = \frac{\left|\mathbf{h}^H\left(\omega,\theta_{\rm s}\right)\mathbf{d}\left(\omega,\theta_{\rm s}\right)\right|^2}{\mathbf{h}^H\left(\omega,\theta_{\rm s}\right)\mathbf{h}\left(\omega,\theta_{\rm s}\right)} = \frac{1}{\mathbf{h}^H\left(\omega,\theta_{\rm s}\right)\mathbf{h}\left(\omega,\theta_{\rm s}\right)}, \tag{2.56}$$

[6] This noise models the sensor noise.

where in the second line of (2.56), the distortionless constraint is assumed. With the delay-and-sum beamformer:

$$\mathbf{h}_{\mathrm{DS}}(\omega, \theta_{\mathrm{s}}) = \frac{\mathbf{d}(\omega, \theta_{\mathrm{s}})}{M}, \tag{2.57}$$

we find the maximum possible white noise gain, which is

$$\mathcal{G}_{\mathrm{wn,max}}(\omega) = M. \tag{2.58}$$

In general, the white noise gain of an $N$th-order CDMA is

$$\mathcal{G}_{\mathrm{wn},N}[\mathbf{h}(\omega, \theta_{\mathrm{s}})] = \frac{1}{\mathbf{h}^H(\omega, \theta_{\mathrm{s}})\mathbf{h}(\omega, \theta_{\mathrm{s}})} \leq M. \tag{2.59}$$

We will see how the white noise may be amplified by CDMAs, especially in low frequencies.

- The diffuse noise[7]:

$$\begin{aligned}[\mathbf{\Gamma}_{\mathbf{v}}(\omega)]_{ij} &= [\mathbf{\Gamma}_{\mathrm{dn}}(\omega)]_{ij} \\ &= \mathrm{sinc}\left(\frac{\omega\delta_{ij}}{c}\right),\end{aligned} \tag{2.60}$$

where

$$\mathrm{sinc}(x) = \frac{\sin x}{x} \tag{2.61}$$

and

$$\delta_{ij} = 2r\left|\sin\left[\frac{\pi(i-j)}{M}\right]\right| \tag{2.62}$$

is the distance between microphones $i$ and $j$. In this scenario, the gain in SNR, $\mathcal{G}_{\mathrm{dn}}[\mathbf{h}(\omega, \theta_{\mathrm{s}})]$, is called the directivity factor and the directivity index is simply defined as [2], [4]

$$\mathcal{D}[\mathbf{h}(\omega, \theta_{\mathrm{s}})] = 10\log_{10}\mathcal{G}_{\mathrm{dn}}[\mathbf{h}(\omega, \theta_{\mathrm{s}})]. \tag{2.63}$$

With diffuse noise, the filter $\mathbf{h}(\omega, \theta_{\mathrm{s}})$ is often found by maximizing the directivity factor. As a result, the optimal filter is given in (2.54) by simply replacing $\mathbf{\Gamma}_{\mathbf{v}}(\omega)$ with $\mathbf{\Gamma}_{\mathrm{dn}}(\omega)$. For a ULA, this filter corresponds to the hypercardioid of order $M-1$, but not for a UCA. We will get back to this point in Chapter 6.

- The noise comes from a point source at the angle $\theta_{\mathrm{n}}$. In this case, the pseudo-coherence matrix is

[7] This situation corresponds to the spherically isotropic noise field.

$$\mathbf{\Gamma}_{\mathrm{v}}(\omega) = \mathbf{d}(\omega, \theta_{\mathrm{n}}) \mathbf{d}^H(\omega, \theta_{\mathrm{n}}), \tag{2.64}$$

where $\mathbf{d}(\omega, \theta_{\mathrm{n}})$ is the steering vector of the noise source. We observe from (2.64) that the pseudo-coherence matrix is singular. In fact, this is the only possibility where the gain in SNR, $\mathcal{G}_{\mathrm{ns}}[\mathbf{h}(\omega, \theta_{\mathrm{s}})]$, is not upper bounded and can go to infinity. We deduce that this gain is

$$\begin{aligned} \mathcal{G}_{\mathrm{ns}}[\mathbf{h}(\omega, \theta_{\mathrm{s}})] &= \frac{\left|\mathbf{h}^H(\omega, \theta_{\mathrm{s}}) \mathbf{d}(\omega, \theta_{\mathrm{s}})\right|^2}{\left|\mathbf{h}^H(\omega, \theta_{\mathrm{s}}) \mathbf{d}(\omega, \theta_{\mathrm{n}})\right|^2} \\ &= \frac{1}{\left|\mathbf{h}^H(\omega, \theta_{\mathrm{s}}) \mathbf{d}(\omega, \theta_{\mathrm{n}})\right|^2}. \end{aligned} \tag{2.65}$$

When the noise and desired signals come from the same direction, i.e., when $\theta_{\mathrm{n}} = \theta_{\mathrm{s}}$, then there is no possible gain, i.e., $\mathcal{G}_{\mathrm{ns}}[\mathbf{h}(\omega, \theta_{\mathrm{s}})] = 1,\ \forall \mathbf{h}(\omega, \theta_{\mathrm{s}})$. We also deduce the gain of an $N$th-order DMA:

$$\mathcal{G}_{\mathrm{ns},N}(\theta_{\mathrm{n}} - \theta_{\mathrm{s}}) = \frac{1}{\left|\mathcal{B}_N(\theta_{\mathrm{n}} - \theta_{\mathrm{s}})\right|^2}. \tag{2.66}$$

The gain $\mathcal{G}_{\mathrm{ns}}[\mathbf{h}(\omega, \theta_{\mathrm{s}})]$ or $\mathcal{G}_{\mathrm{ns},N}(\theta_{\mathrm{n}} - \theta_{\mathrm{s}})$ depends only on the beampattern; so the plot of this gain as a function of $\theta_{\mathrm{n}}$ is equivalent to the polar pattern.

We now give an important property concerning $\mathbf{\Gamma}_{\mathrm{dn}}(\omega)$.

*Property 2.4.* The diffuse noise pseudo-coherence matrix is a circulant matrix.

*Proof.* We recall that, by its own definition, $\mathbf{\Gamma}_{\mathrm{dn}}(\omega)$ is symmetric and Toeplitz. A Toeplitz matrix is circulant if and only if the elements of $\mathbf{\Gamma}_{\mathrm{dn}}(\omega)$ at the lines $1, 2, \ldots, M-1$ and last column are equal to the elements at the lines $2, 3, \ldots, M$ and first column, i.e.,

$$[\mathbf{\Gamma}_{\mathrm{dn}}(\omega)]_{mM} = [\mathbf{\Gamma}_{\mathrm{dn}}(\omega)]_{(m+1)1}, \ m = 1, 2, \ldots, M-1. \tag{2.67}$$

This is true if we can show that

$$\delta_{mM} = \delta_{(m+1)1}, \ m = 1, 2, \ldots, M-1. \tag{2.68}$$

This is, indeed, straightforward to verify by using the trigonometric identity:

$$\sin(\alpha + \beta) = \sin\alpha\cos\beta + \cos\alpha\sin\beta. \tag{2.69}$$

## References

1. R. A. Monzingo and T. W. Miller, *Introduction to Adaptive Arrays.* Raleigh, NC: SciTech, 2004.

2. G. W. Elko and J. Meyer, "Microphone arrays," in *Springer Handbook of Speech Processing*, J. Benesty, M. M. Sondhi, and Y. Huang, Eds. Berlin, Germany: Springer-Verlag, 2008, Chapter 50, Part I, pp. 1021–1041.
3. M. Buck, "Aspects of first-order differential microphone arrays in the presence of sensor imperfections," *European Trans. Telecommunications*, vol. 13, pp. 115-122, Mar.-Apr. 2002.
4. G. W. Elko, "Superdirectional microphone arrays," in *Acoustic Signal Processing for Telecommunication*, S. L. Gay and J. Benesty, Eds. Boston, MA: Kluwer Academic Publishers, 2000, Chapter 10, pp. 181–237.
5. J. Benesty and J. Chen, *Study and Design of Differential Microphone Arrays.* Berlin, Germany: Springer-Verlag, 2012.
6. J. Benesty, J. Chen, and E. Habets, *Speech Enhancement in the STFT Domain.* Springer Briefs in Electrical and Computer Engineering, 2011.
7. M. Buck and M. Rößler, "First order differential microphone arrays for automative applications," in *Proc. IWAENC*, 2001.
8. E. De Sena, H. Hacihabiboğlu, and Z. Cvetković, "On the design and implementation of higher-order differential microphones," *IEEE Trans. Audio, Speech, Language Process.*, vol. 20, pp. 162–174, Jan. 2012.
9. R. N. Marshall and W. R. Harry, "A new microphone providing uniform directivity over an extended frequency range," *J. Acoust. Soc. Am.*, vol. 12, pp. 481–497, 1941.

# Chapter 3
# Design of First-Order Circular Differential Arrays

This chapter is dedicated to the design of first-order circular differential microphone arrays with three sensors. We explore the dipole, cardioid, subcardioid, hypercardioid, and supercardioid.

## 3.1 Principle

In this work, the proposed approach to design CDMAs is based on the obvious observation that any ideal frequency-independent directivity pattern has a one at the angle $\theta = \theta_{\text{s}}$ and a number of nulls in some specific directions.

First-order CDMAs can be designed with two or three microphones. The case $M = 2$ has no interest in this investigation since it corresponds to an LDMA, which has been widely studied in the literature [1], [2]. Therefore, our focus is on the case $M = 3$. The geometry of this CDMA is, in fact, an equilateral triangle whose circumscribed circle is what we have in mind when we refer to a UCA with three sensors. The circumradius is $r$ and the sides length of the triangle (or intersensor spacing) is $\delta = 2r \sin(\pi/3) = \sqrt{3}r$. In this situation, the positions of the three microphones are

$$\psi_1 = 0,\ \psi_2 = \frac{2\pi}{3},\ \psi_3 = \frac{4\pi}{3}, \tag{3.1}$$

and we can perfectly steer at the three different angles: $\theta_{\text{s}} = \psi_m,\ m = 1, 2, 3$. In all the rest, we choose $\theta_{\text{s}} = 0$. Since we have three microphones, we have exactly three constraints to fulfill. The first one is the distortionless response (a one at the angle $\theta = 0$), the second constraint is a null at $\theta = \theta_{1,1}$ with $0 < \theta_{1,1} \leq \pi$, and the last one is on the symmetry of the beampattern, i.e., $H_2(\omega) = H_3(\omega)$. Thus, these three constraints can be written as



J. Benesty et al., *Design of Circular Differential Microphone Arrays*,
Springer Topics in Signal Processing 12, DOI 10.1007/978-3-319-14842-7_3

$$\mathbf{d}^H(\omega, 0)\,\mathbf{h}(\omega) = 1, \quad (3.2)$$

$$\mathbf{d}^H(\omega, \theta_{1,1})\,\mathbf{h}(\omega) = 0, \quad (3.3)$$

$$\mathbf{c}_{3,1}^T\mathbf{h}(\omega) = 0, \quad (3.4)$$

where

$$\mathbf{c}_{3,1} = \begin{bmatrix} 0 \ 1 \ -1 \end{bmatrix}^T \quad (3.5)$$

and $\theta_{1,1}$ is given by design (a null at the angle $\theta_{1,1}$). We can express (3.2)–(3.4) as

$$\mathbf{A}_3(\omega, \theta)\,\mathbf{h}(\omega) = \mathbf{b}_3, \quad (3.6)$$

where

$$\mathbf{A}_3(\omega, \theta) = \begin{bmatrix} \mathbf{d}^H(\omega, 0) \\ \mathbf{d}^H(\omega, \theta_{1,1}) \\ \mathbf{c}_{3,1}^T \end{bmatrix} = \begin{bmatrix} e^{-\jmath\varpi} & e^{-\jmath\varpi\cos\psi_2} & e^{-\jmath\varpi\cos\psi_3} \\ e^{-\jmath\varpi\cos\theta_{1,1}} & e^{-\jmath\varpi\cos(\theta_{1,1}-\psi_2)} & e^{-\jmath\varpi\cos(\theta_{1,1}-\psi_3)} \\ 0 & 1 & -1 \end{bmatrix} \quad (3.7)$$

and

$$\mathbf{b}_3 = \begin{bmatrix} 1 \ 0 \ 0 \end{bmatrix}^T. \quad (3.8)$$

It is clear that $\mathbf{A}_3(\omega, \theta)$ is a full-rank square matrix as long as $\theta_{1,1} \neq 0$, which was already assumed. As a result, (3.6) has a unique solution and the designed filter is given by

$$\mathbf{h}(\omega) = \mathbf{A}_3^{-1}(\omega, \theta)\,\mathbf{b}_3. \quad (3.9)$$

Once the first-order CDMA filter is derived with three microphones, it is easy to compute the beampattern:

$$|\mathcal{B}_1[\mathbf{h}(\omega), \theta]|^2 = \left|\mathbf{d}^H(\omega, \theta)\,\mathbf{A}_3^{-1}(\omega, \theta)\,\mathbf{b}_3\right|^2, \quad (3.10)$$

the directivity factor:

$$\mathcal{G}_{\mathrm{dn},1}[\mathbf{h}(\omega)] = \frac{1}{\mathbf{b}_3^T\mathbf{A}_3^{-H}(\omega, \theta)\,\boldsymbol{\Gamma}_{\mathrm{dn}}(\omega)\,\mathbf{A}_3^{-1}(\omega, \theta)\,\mathbf{b}_3}, \quad (3.11)$$

and the white noise gain:

$$\mathcal{G}_{\mathrm{wn},1}[\mathbf{h}(\omega)] = \frac{1}{\mathbf{b}_3^T\mathbf{A}_3^{-H}(\omega, \theta)\,\mathbf{A}_3^{-1}(\omega, \theta)\,\mathbf{b}_3}. \quad (3.12)$$

The white noise gain depends in a large part on the condition number [3] of $\mathbf{A}_3(\omega, \theta)$; the higher is this condition number the more the white noise amplification [2]. Also, the closer is $\theta_{1,1}$ to 0, the larger is the condition number of $\mathbf{A}_3(\omega, \theta)$. The lowest condition number is obtained when $\theta_{1,1} = \pi$.

There is an alternative way to design the filter $\mathbf{h}(\omega)$. Since $H_2(\omega) = H_3(\omega)$, we can restrict ourself to design the filter of length 2:

$$\mathbf{h}'(\omega) = \left[ H_1(\omega) \; H_2(\omega) \right]^T. \tag{3.13}$$

Then, it is not hard to find that

$$\mathbf{A}'_2(\omega, \theta) \mathbf{h}'(\omega) = \mathbf{b}'_2, \tag{3.14}$$

where

$$\mathbf{A}'_2(\omega, \theta) = \begin{bmatrix} \mathbf{d}'^H(\omega, 0) \\ \mathbf{d}'^H(\omega, \theta_{1,1}) \end{bmatrix} \tag{3.15}$$
$$= \begin{bmatrix} e^{-\jmath\varpi} & e^{-\jmath\varpi\cos\psi_2} + e^{-\jmath\varpi\cos\psi_3} \\ e^{-\jmath\varpi\cos\theta_{1,1}} & e^{-\jmath\varpi\cos(\theta_{1,1}-\psi_2)} + e^{-\jmath\varpi\cos(\theta_{1,1}-\psi_3)} \end{bmatrix}$$

and

$$\mathbf{b}'_2 = \begin{bmatrix} 1 \\ 0 \end{bmatrix}. \tag{3.16}$$

We deduce that

$$\mathbf{h}'(\omega) = \mathbf{A}'^{-1}_2(\omega, \theta) \mathbf{b}'_2 \tag{3.17}$$

and

$$\left|\mathcal{B}_1\left[\mathbf{h}'(\omega), \theta\right]\right|^2 = \left|\mathbf{d}'^H(\omega, \theta) \mathbf{A}'^{-1}_2(\omega, \theta) \mathbf{b}'_2\right|^2 \tag{3.18}$$
$$= \left|\mathcal{B}_1\left[\mathbf{h}(\omega), \theta\right]\right|^2,$$

where

$$\mathbf{d}'(\omega, \theta) = \left[ e^{\jmath\varpi\cos\theta} \; e^{\jmath\varpi\cos(\theta-\psi_2)} + e^{\jmath\varpi\cos(\theta-\psi_3)} \right]^T, \tag{3.19}$$

with $\mathbf{d}'(\omega, \theta) = \mathbf{d}'(\omega, -\theta)$.

The output of the proposed beamformer with a three-sensor UCA can be seen as the sum of two beamformers' outputs with two-sensor ULAs. One ULA is composed of microphones 1 and 2, and the other one is composed of microphones 1 and 3. For both ULAs, the output of sensor 1 is weighted by 1/2.

The steering vector, $\mathbf{d}'(\omega, \theta)$, can be rewritten as

$$\begin{aligned}\mathbf{d}'(\omega,\theta) &= \left[e^{\jmath\varpi\cos\theta}\; e^{\jmath\varpi\cos(\theta-\psi_2)} + e^{\jmath\varpi\cos(\theta+\psi_2)}\right]^T \\ &= \left[e^{\jmath\varpi\cos\theta}\; 2\cos\left(\tfrac{\sqrt{3}}{2}\varpi\sin\theta\right) e^{-\jmath\varpi\cos\theta}\right]^T \\ &= e^{\jmath\varpi\cos\theta}\left[1\; 2\cos\left(\tfrac{\sqrt{3}}{2}\varpi\sin\theta\right) e^{-\jmath 2\varpi\cos\theta}\right]^T.\end{aligned} \tag{3.20}$$

Therefore, the proposed three-sensor CDMA can be seen as a two-sensor non-isotropic LDMA for which the interelement spacing is

$$\delta' = 2r = \frac{2}{\sqrt{3}}\delta. \tag{3.21}$$

## 3.2 Design Examples

In this section, important particular cases of first-order CDMAs with three sensors are numerically studied.

### *3.2.1 First-Order Dipole*

The first-order dipole has a one at the angle $\theta_s = 0$ and a null at the angle $\theta_s + \pi/2$. With these values, i.e., $\theta_s = 0$ and $\theta_{1,1} = \pi/2$, it is easy to determine the filter corresponding to the first-order dipole from (3.9).

Figure 3.1 displays the patterns of the first-order dipole with three microphones for several frequencies and two values of $\delta$ (1 and 2 cm). Clearly, the designed patterns are identical to the desired pattern shown in Fig. 2.2(a).

Figure 3.2 gives plots of $\mathcal{G}_{\mathrm{dn},1}[\mathbf{h}(\omega)]$ [where $\mathbf{h}(\omega)$ corresponds to the first-order dipole filter with three microphones], as a function of frequency, for different values of $\delta$. For $\delta$ small, this directivity factor is constant across frequencies; when $\delta$ increases, the directivity factor decreases, especially in high frequencies.

Figure 3.3 gives plots of $\mathcal{G}_{\mathrm{wn},1}[\mathbf{h}(\omega)]$ [where $\mathbf{h}(\omega)$ corresponds to the first-order dipole filter with three microphones], as a function of frequency, for different values of $\delta$. As expected, this white noise gain improves when the interelement spacing increases. We can even have positive gains in high frequencies

From all these simulations, we can say that with a UCA of three microphones and $\delta = 3$ cm, we can get a very nice first-order dipole. Moreover, this UCA can perfectly steer in three different directions.

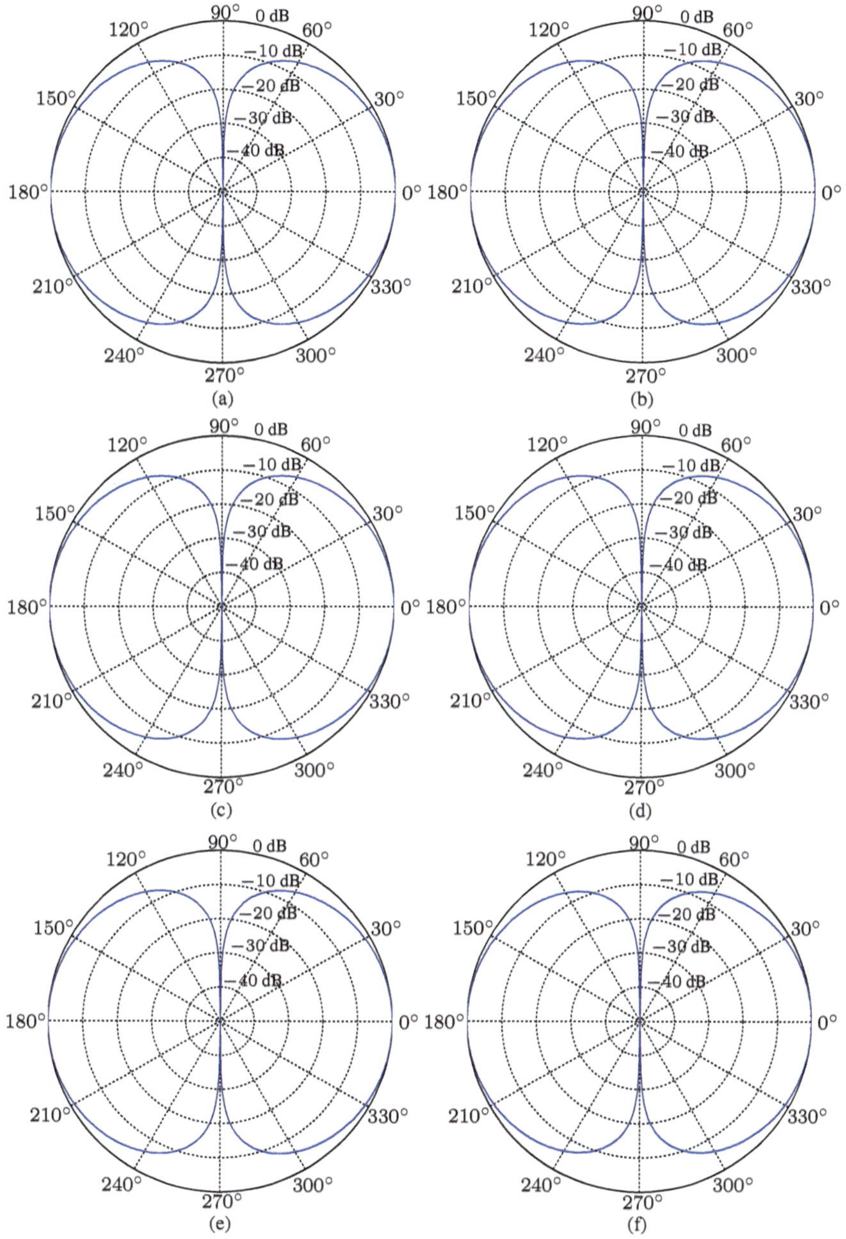


**Fig. 3.1** Patterns of the first-order dipole (with three microphones) for several frequencies and two values of $\delta$: (a) $f = 0.5$ kHz, $\delta = 1$ cm, (b) $f = 0.5$ kHz, $\delta = 2$ cm, (c) $f = 1$ kHz, $\delta = 1$ cm, (d) $f = 1$ kHz, $\delta = 2$ cm, (e) $f = 3$ kHz, $\delta = 1$ cm, and (f) $f = 3$ kHz, $\delta = 2$ cm.

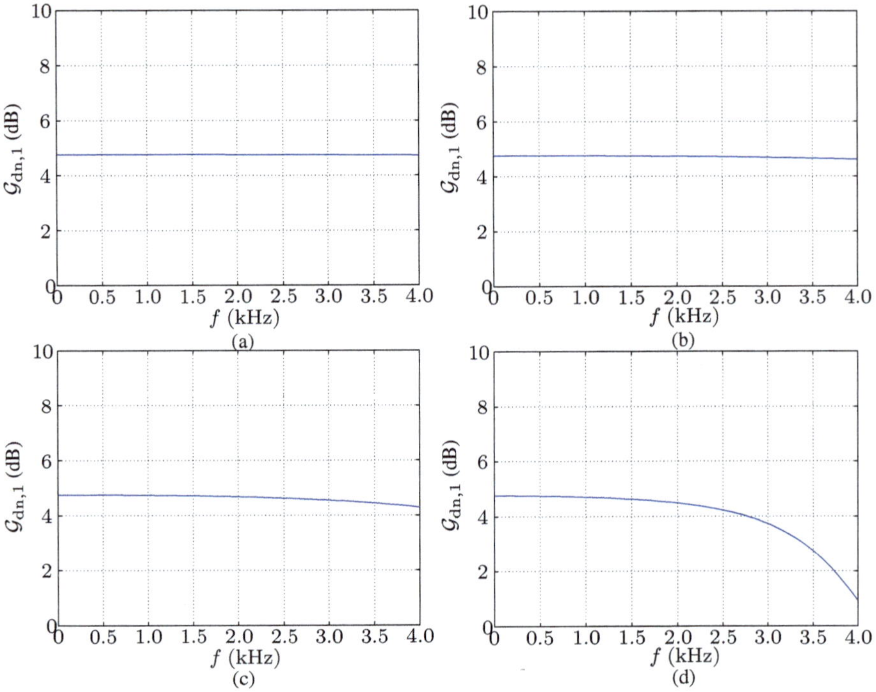


**Fig. 3.2** The directivity factor of the first-order dipole (with three microphones), as a function of frequency, for different values of $\delta$: (a) $\delta = 1$ cm, (b) $\delta = 2$ cm, (c) $\delta = 3$ cm, and (d) $\delta = 5$ cm.

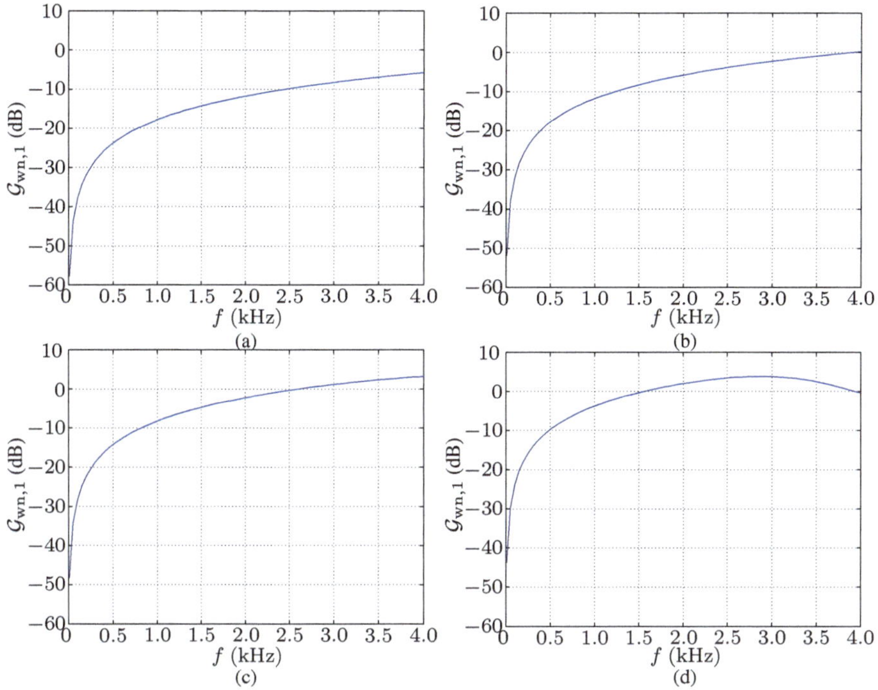


**Fig. 3.3** The white noise gain of the first-order dipole (with three microphones), as a function of frequency, for different values of $\delta$: (a) $\delta = 1$ cm, (b) $\delta = 2$ cm, (c) $\delta = 3$ cm, and (d) $\delta = 5$ cm.

### 3.2.2 First-Order Cardioid

In the first-order cardioid, there is a one at the angle $\theta_s = 0$ and a null at the angle $\theta_s + \pi$. We see that $\theta_{1,1} = \pi$ and the filter corresponding to the first-order cardioid is easily obtained from (3.9).

Figure 3.4 shows the patterns of the first-order cardioid with three microphones for several frequencies and two values of $\delta$ (1 and 2 cm). These patterns are identical to the ideal pattern given in Fig. 2.2(b).

Figure 3.5 gives plots of $\mathcal{G}_{dn,1}[\mathbf{h}(\omega)]$ [where $\mathbf{h}(\omega)$ corresponds to the first-order cardioid filter with three microphones], as a function of frequency, for different values of $\delta$. This directivity factor is close to 5 dB for small values of $\delta$ but it decreases quite quickly in high frequencies when $\delta$ increases.

Figure 3.6 displays plots of $\mathcal{G}_{wn,1}[\mathbf{h}(\omega)]$ [where $\mathbf{h}(\omega)$ corresponds to the first-order cardioid filter with three microphones], as a function of frequency, for different values of $\delta$. For $\delta = 2$ cm, this white noise gain is positive above 2 kHz, while its value is around $-10$ dB around 500 Hz, which is not that bad.

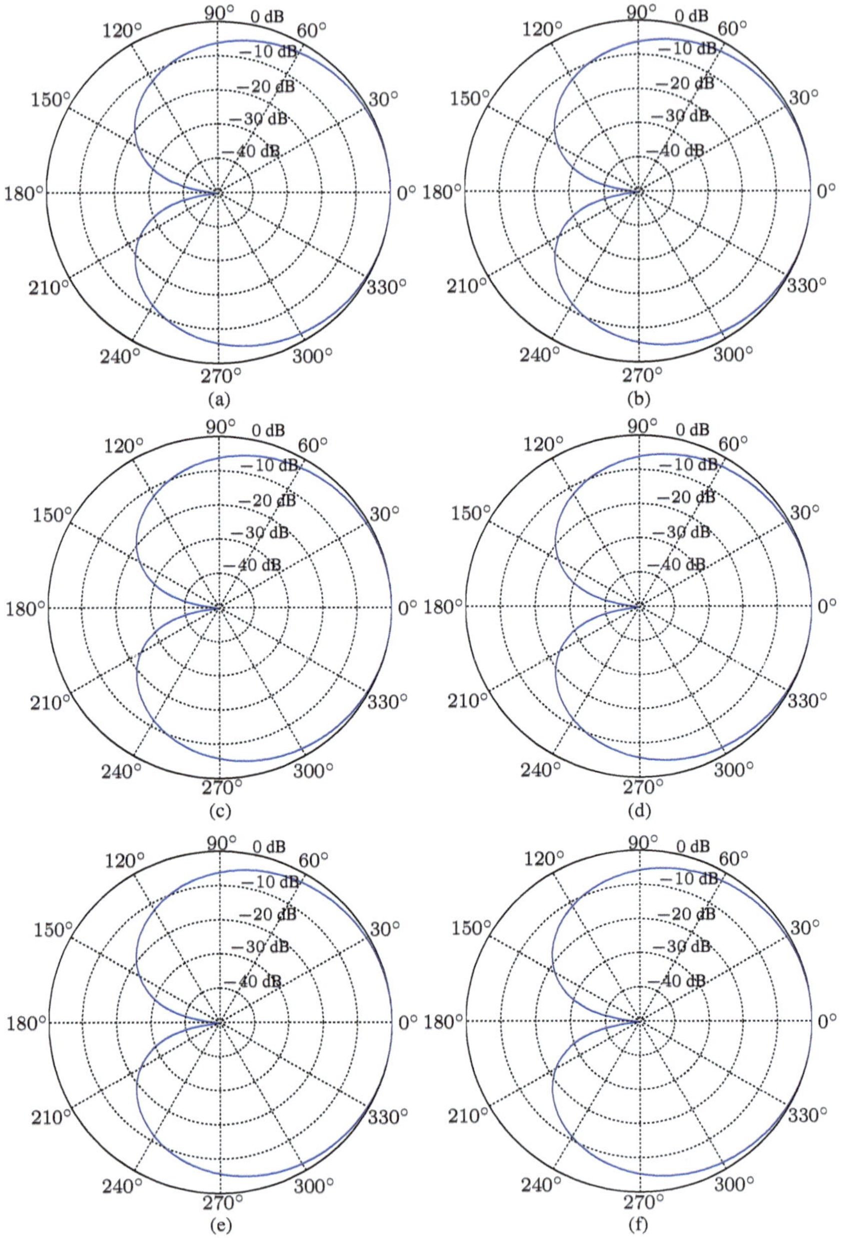


**Fig. 3.4** Patterns of the first-order cardioid (with three microphones) for several frequencies and two values of $\delta$: (a) $f = 0.5$ kHz, $\delta = 1$ cm, (b) $f = 0.5$ kHz, $\delta = 2$ cm, (c) $f = 1$ kHz, $\delta = 1$ cm, (d) $f = 1$ kHz, $\delta = 2$ cm, (e) $f = 3$ kHz, $\delta = 1$ cm, and (f) $f = 3$ kHz, $\delta = 2$ cm.

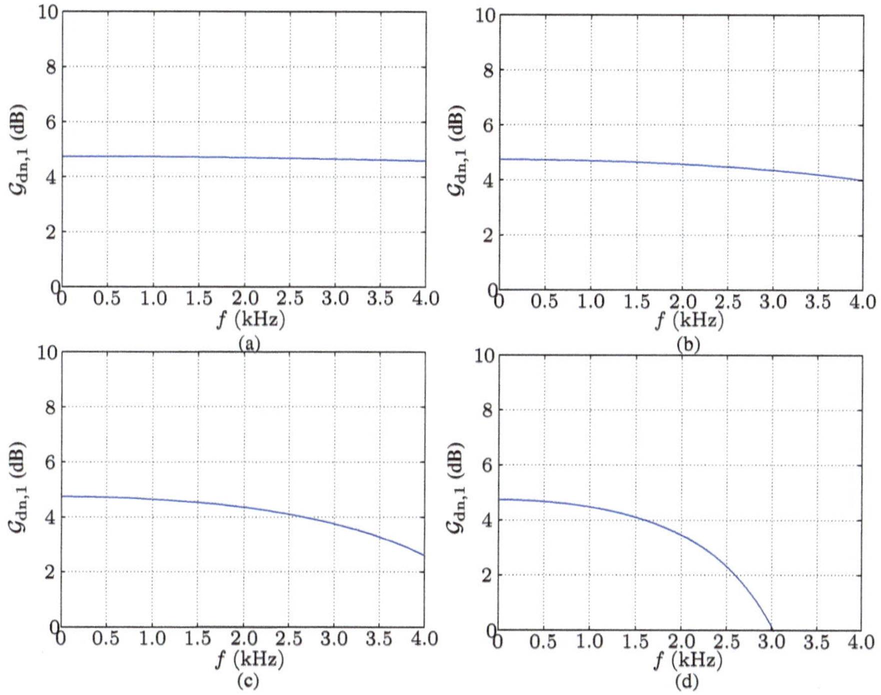


**Fig. 3.5** The directivity factor of the first-order cardioid (with three microphones), as a function of frequency, for different values of $\delta$: (a) $\delta = 1$ cm, (b) $\delta = 2$ cm, (c) $\delta = 3$ cm, and (d) $\delta = 5$ cm.

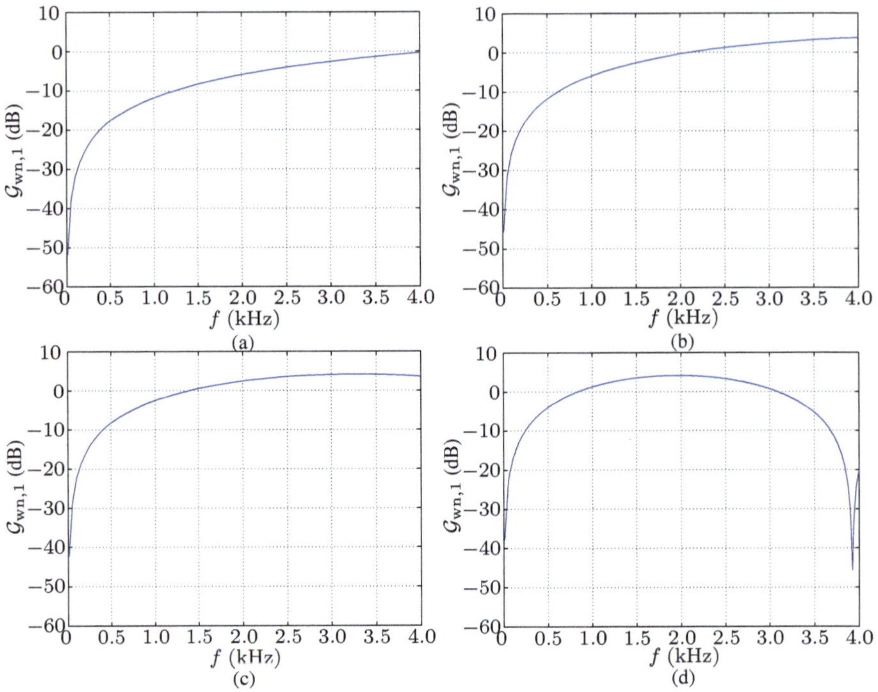


**Fig. 3.6** The white noise gain of the first-order cardioid (with three microphones), as a function of frequency, for different values of $\delta$: (a) $\delta = 1$ cm, (b) $\delta = 2$ cm, (c) $\delta = 3$ cm, and (d) $\delta = 5$ cm.

We conclude this subsection by saying that a very good first-order cardioid can be designed with a 3-element UCA and $\delta = 2$ cm, which can steer in three different directions.

### *3.2.3 General First-Order Circular Differential Array*

We consider designing any first-order circular differential array with three sensors from the linear system of three equations:

$$\begin{bmatrix} \mathbf{d}^H(\omega, 0) \\ \mathbf{d}^H(\omega, \theta_{1,1}) \\ \mathbf{c}_{3,1}^T \end{bmatrix} \mathbf{h}(\omega) = \begin{bmatrix} 1 \\ \beta_{1,1} \\ 0 \end{bmatrix}, \tag{3.22}$$

where $0 < \theta_{1,1} \leq \pi$ and $0 \leq \beta_{1,1} \leq 1$. We immediately find that the unique solution to (3.22) is

$$\mathbf{h}(\omega) = \mathbf{A}_3^{-1}(\omega, \theta) \begin{bmatrix} 1 \\ \beta_{1,1} \\ 0 \end{bmatrix}. \tag{3.23}$$

Alternatively, we can solve, instead,

$$\mathbf{h}'(\omega) = \mathbf{A}_2'^{-1}(\omega, \theta) \begin{bmatrix} 1 \\ \beta_{1,1} \end{bmatrix}, \tag{3.24}$$

where $\mathbf{A}_2'(\omega, \theta)$ is defined in the previous section.

Depending on the values of $\theta_{1,1}$ and $\beta_{1,1}$, we find five interesting first-order CDMA filters.

- Dipole: $\theta_{1,1} = \pi/2,\ \beta_{1,1} = 0$ (see Subsection 3.2.1).
- Cardioid: $\theta_{1,1} = \pi,\ \beta_{1,1} = 0$ (see Subsection 3.2.2).
- Subcardioid [4], [5]: $\theta_{1,1} = \pi,\ \beta_{1,1} = 0.4$.
- Hypercardioid: $\theta_{1,1} = 2\pi/3,\ \beta_{1,1} = 0$.
- Supercardioid: $\theta_{1,1} = 3\pi/4,\ \beta_{1,1} = 0$.

Figures 3.7, 3.8, and 3.9 display the patterns [with $\mathbf{h}(\omega)$ defined in (3.23)] of the subcardioid, hypercardioid, and supercardioid, respectively, for several frequencies and two values of $\delta$ (1 and 2 cm). Again, we can verify that all the obtained patterns are identical to the desired ones.

Figures 3.10, 3.11, and 3.12 give plots of the directivity factor, as a function of frequency, for the subcardioid, hypercardioid, and supercardioid, respectively, for different values of $\delta$.

Figures 3.13, 3.14, and 3.15 show plots of the white noise gain, as a function of frequency, for the subcardioid, hypercardioid, and supercardioid, respectively, for different values of $\delta$.

For the subcardioid, we can get a directivity factor close to 3 dB while it is almost 6 dB for the hypercardioid and supercardioid. A good value for the interelement spacing is 2 cm.

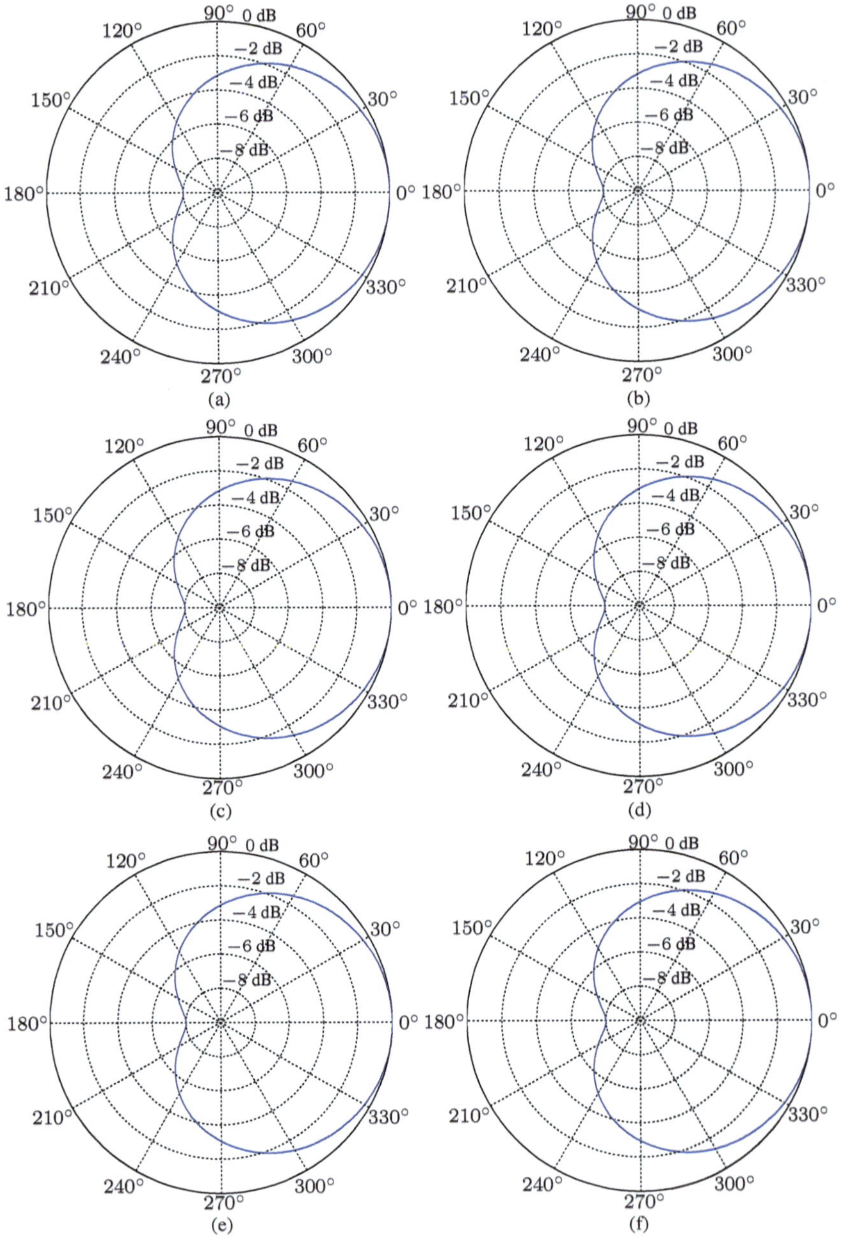


**Fig. 3.7** Patterns of the first-order subcardioid (with three microphones) for several frequencies and two values of $\delta$: (a) $f = 0.5$ kHz, $\delta = 1$ cm, (b) $f = 0.5$ kHz, $\delta = 2$ cm, (c) $f = 1$ kHz, $\delta = 1$ cm, (d) $f = 1$ kHz, $\delta = 2$ cm, (e) $f = 3$ kHz, $\delta = 1$ cm, and (f) $f = 3$ kHz, $\delta = 2$ cm.

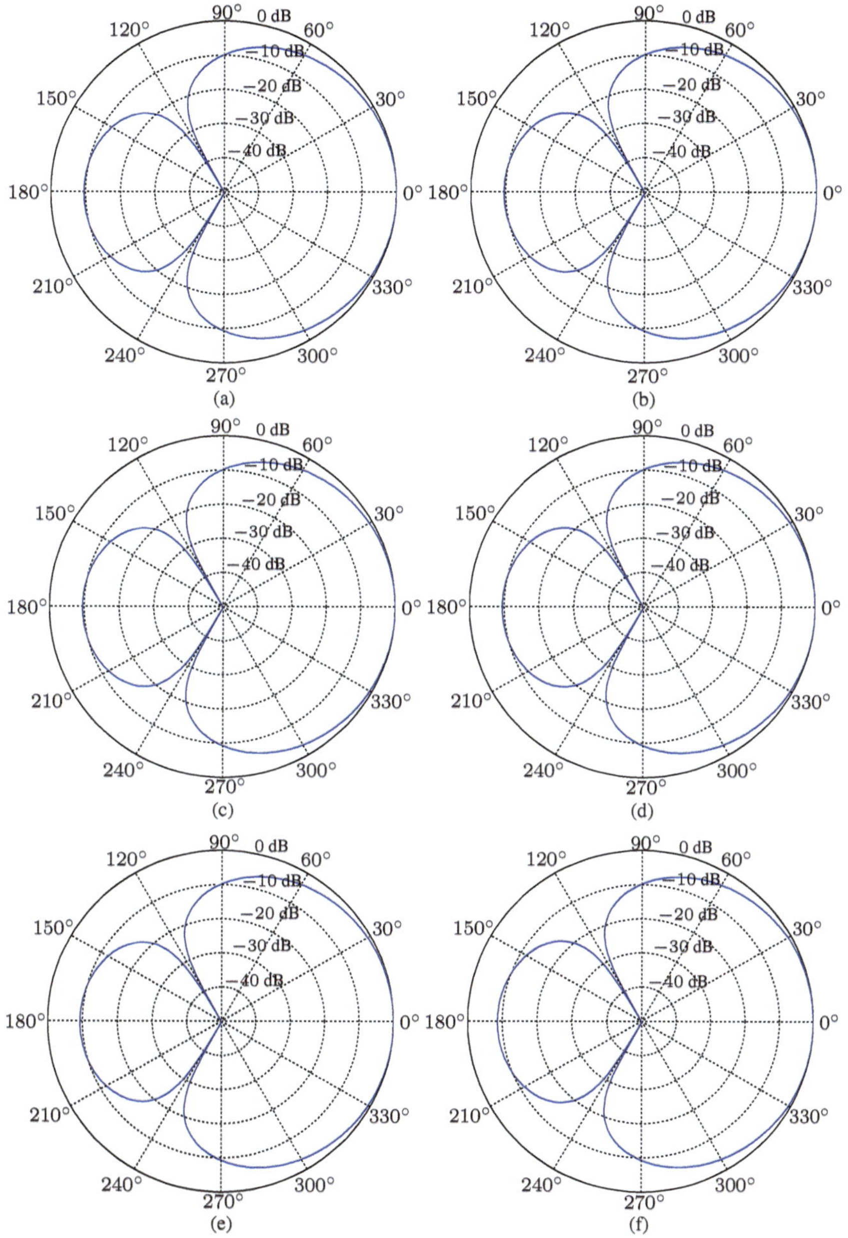


**Fig. 3.8** Patterns of the first-order hypercardioid (with three microphones) for several frequencies and two values of $\delta$: (a) $f = 0.5$ kHz, $\delta = 1$ cm, (b) $f = 0.5$ kHz, $\delta = 2$ cm, (c) $f = 1$ kHz, $\delta = 1$ cm, (d) $f = 1$ kHz, $\delta = 2$ cm, (e) $f = 3$ kHz, $\delta = 1$ cm, and (f) $f = 3$ kHz, $\delta = 2$ cm.

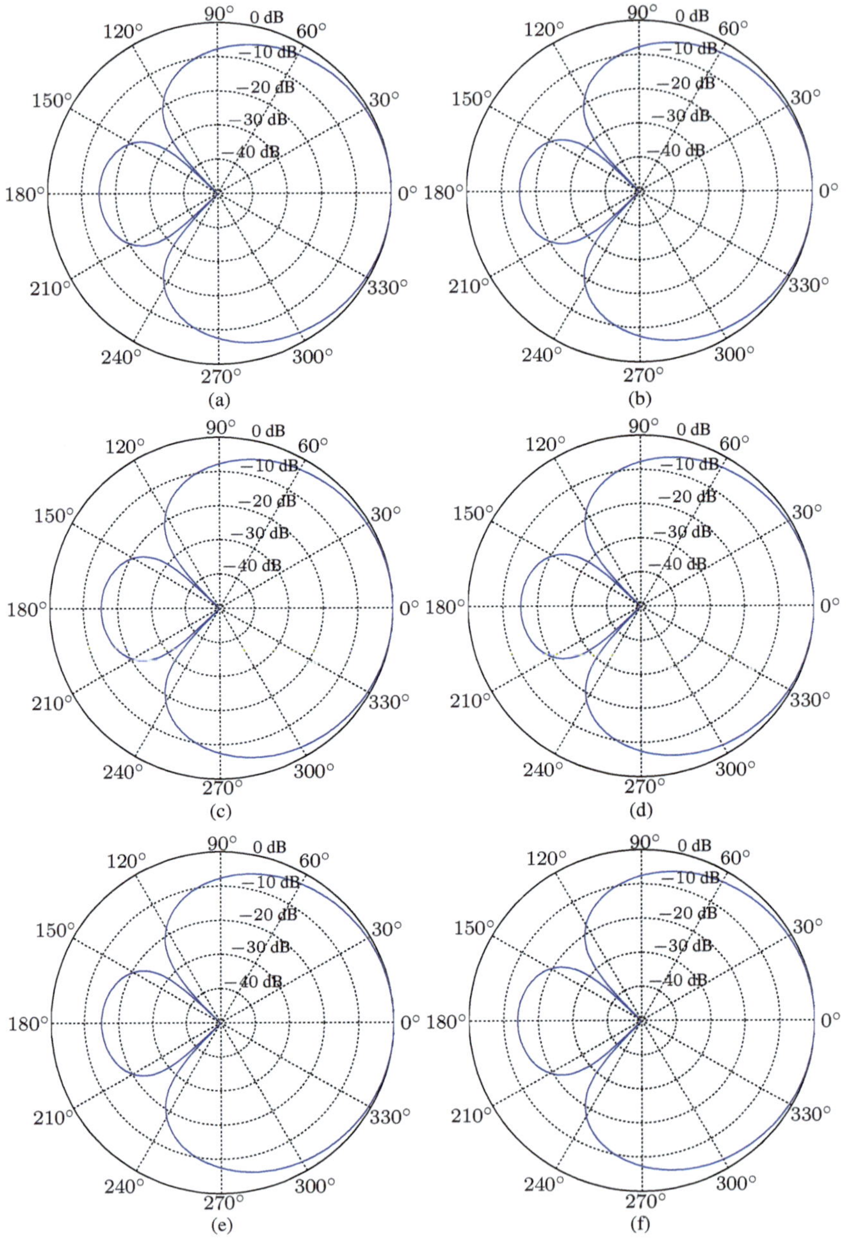


**Fig. 3.9** Patterns of the first-order supercardioid (with three microphones) for several frequencies and two values of $\delta$: (a) $f = 0.5$ kHz, $\delta = 1$ cm, (b) $f = 0.5$ kHz, $\delta = 2$ cm, (c) $f = 1$ kHz, $\delta = 1$ cm, (d) $f = 1$ kHz, $\delta = 2$ cm, (e) $f = 3$ kHz, $\delta = 1$ cm, and (f) $f = 3$ kHz, $\delta = 2$ cm.

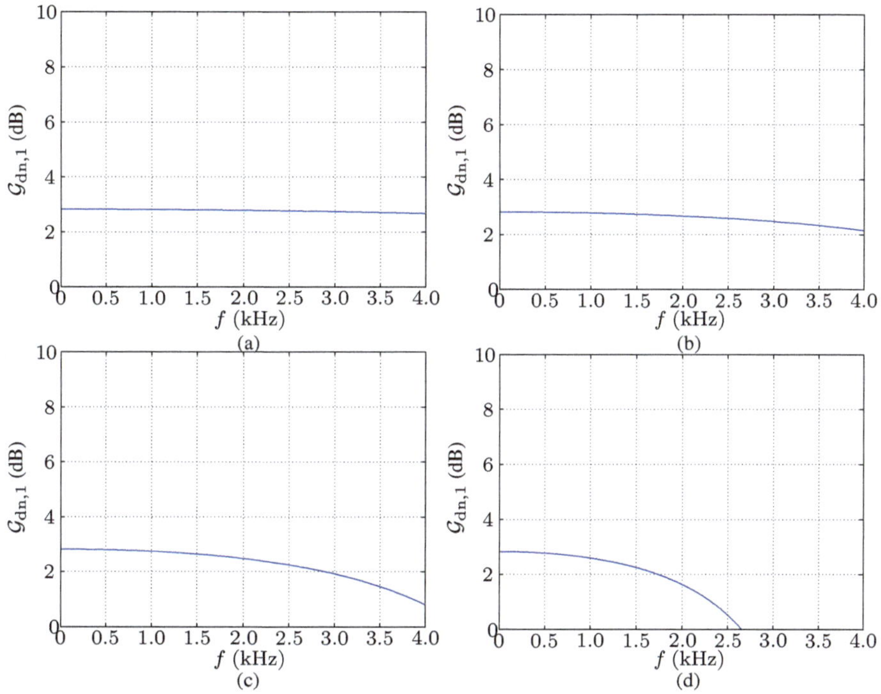


**Fig. 3.10** The directivity factor of the first-order subcardioid (with three microphones), as a function of frequency, for different values of $\delta$: (a) $\delta = 1$ cm, (b) $\delta = 2$ cm, (c) $\delta = 3$ cm, and (d) $\delta = 5$ cm.

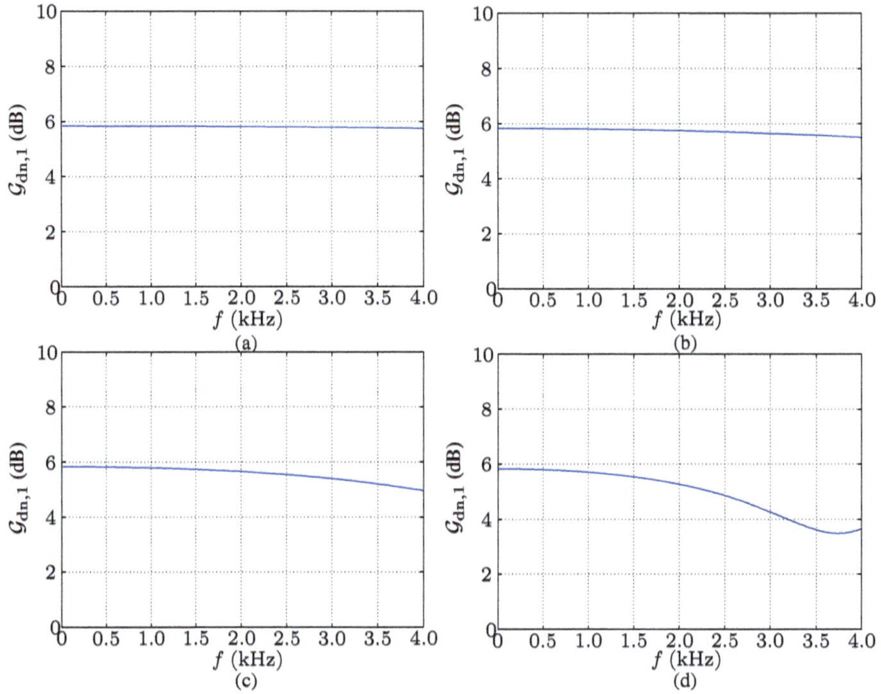


**Fig. 3.11** The directivity factor of the first-order hypercardioid (with three microphones), as a function of frequency, for different values of $\delta$: (a) $\delta = 1$ cm, (b) $\delta = 2$ cm, (c) $\delta = 3$ cm, and (d) $\delta = 5$ cm.

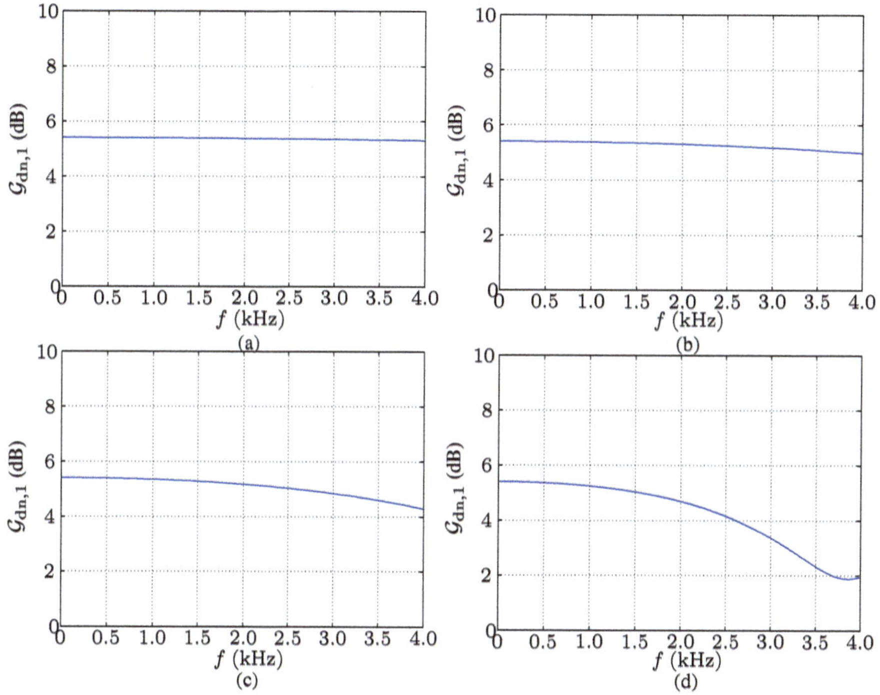


**Fig. 3.12** The directivity factor of the first-order supercardioid (with three microphones), as a function of frequency, for different values of $\delta$: (a) $\delta = 1$ cm, (b) $\delta = 2$ cm, (c) $\delta = 3$ cm, and (d) $\delta = 5$ cm.

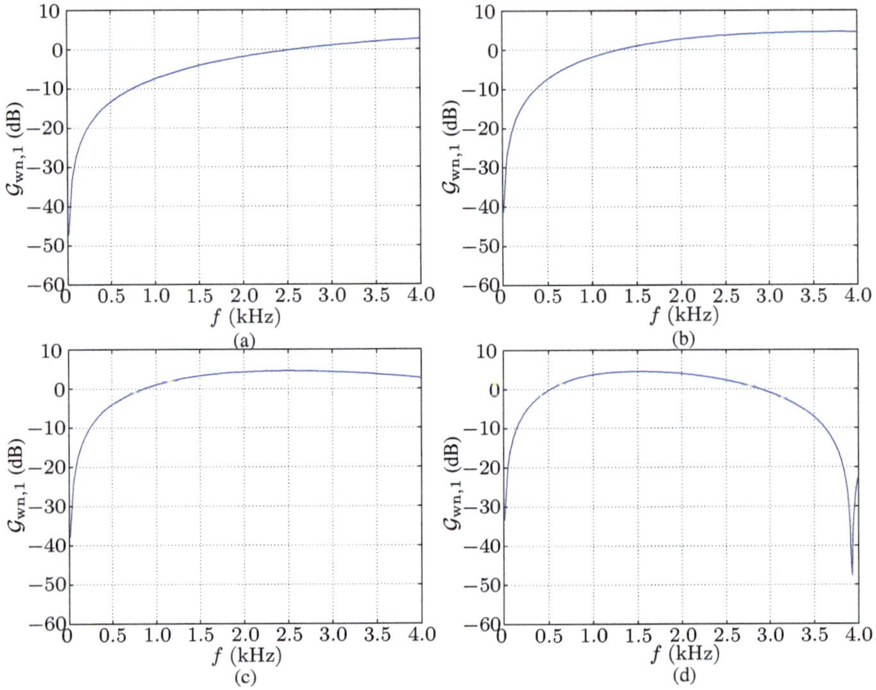


**Fig. 3.13** The white noise gain of the first-order subcardioid (with three microphones), as a function of frequency, for different values of $\delta$: (a) $\delta = 1$ cm, (b) $\delta = 2$ cm, (c) $\delta = 3$ cm, and (d) $\delta = 5$ cm.

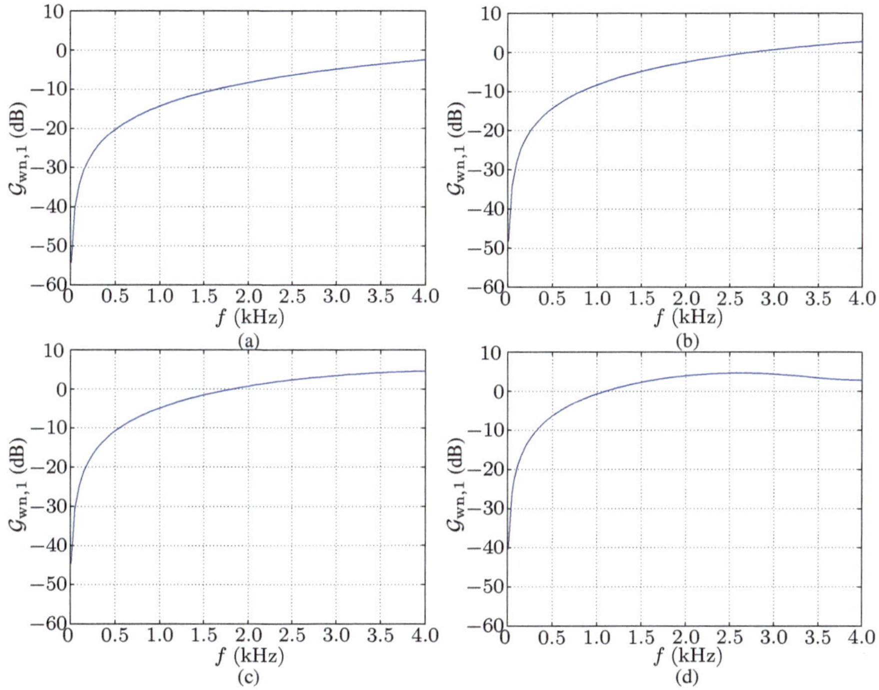


**Fig. 3.14** The white noise gain of the first-order hypercardioid (with three microphones), as a function of frequency, for different values of $\delta$: (a) $\delta = 1$ cm, (b) $\delta = 2$ cm, (c) $\delta = 3$ cm, and (d) $\delta = 5$ cm.

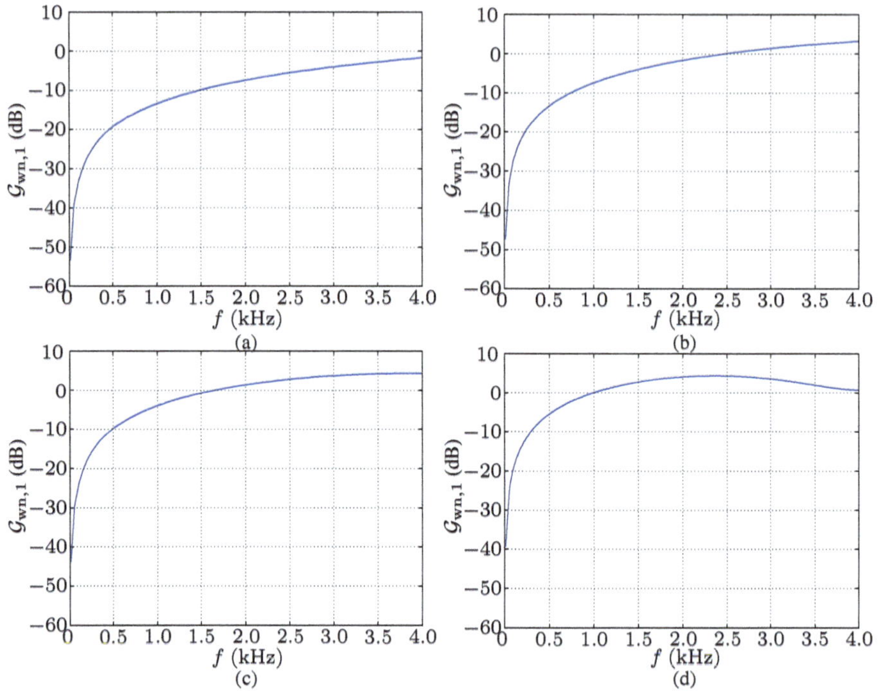


**Fig. 3.15** The white noise gain of the first-order supercardioid (with three microphones), as a function of frequency, for different values of $\delta$: (a) $\delta = 1$ cm, (b) $\delta = 2$ cm, (c) $\delta = 3$ cm, and (d) $\delta = 5$ cm.

From all the extensive simulations presented in this chapter, we can conclude that any first-order directivity pattern can be designed with a 3-element UCA, which can perfectly steer in three different directions.

# References

1. G. W. Elko, "Superdirectional microphone arrays," in *Acoustic Signal Processing for Telecommunication*, S. L. Gay and J. Benesty, Eds. Boston, MA: Kluwer Academic Publishers, 2000, Chapter 10, pp. 181–237.
2. J. Benesty and J. Chen, *Study and Design of Differential Microphone Arrays*. Berlin, Germany: Springer-Verlag, 2012.
3. G. H. Golub and C. F. Van Loan, *Matrix Computations*. Third Edition. Baltimore, Maryland: The Johns Hopkins University Press, 1996.
4. J. Merimaa, "Applications of 3-D microphone array," in *Proc. AES 112th Convention*, 2002, pp. 1–11.
5. E. De Sena, H. Hacihabiboğlu, and Z. Cvetković, "On the design and implementation of higher-order differential microphones," *IEEE Trans. Audio, Speech, Language Process.*, vol. 20, pp. 162–174, Jan. 2012.

# Chapter 4
# Design of Second-Order Circular Differential Arrays

This chapter is dedicated to the design of second-order circular differential microphone arrays with four and five sensors. The most well-known directivity patterns [1], [2] are designed and studied in each case.

## 4.1 Principle and Design Examples with Four Microphones

A second-order CDMA can be designed with four microphones. In this case, the geometry of the array is a square whose circumradius is $r$ and its sides length (or intersensor spacing) is $\delta = 2r \sin(\pi/4) = \sqrt{2}r$. We deduce that the positions of the four microphones are

$$\psi_1 = 0, \ \psi_2 = \frac{\pi}{2}, \ \psi_3 = \pi, \ \psi_4 = \frac{3\pi}{2}, \tag{4.1}$$

and we can perfectly steer at the four different angles: $\theta_s = \psi_m$, $m = 1, 2, 3, 4$. In all the rest, we choose $\theta_s = 0$. Because of the symmetry constraint, we must have $H_2(\omega) = H_4(\omega)$. Then, it is not hard to find that the linear system to solve is

$$\mathbf{A}_4(\omega, \theta)\, \mathbf{h}(\omega) = \mathbf{b}_4, \tag{4.2}$$

where

$$\mathbf{A}_4(\omega, \theta) = \begin{bmatrix} \mathbf{d}^H(\omega, 0) \\ \mathbf{d}^H(\omega, \theta_{2,1}) \\ \mathbf{d}^H(\omega, \theta_{2,2}) \\ \mathbf{c}_{4,1}^T \end{bmatrix}, \tag{4.3}$$



J. Benesty et al., *Design of Circular Differential Microphone Arrays*,
Springer Topics in Signal Processing 12, DOI 10.1007/978-3-319-14842-7_4

$$\mathbf{c}_{4,1} = \begin{bmatrix} 0 \; 1 \; 0 \; -1 \end{bmatrix}^T, \tag{4.4}$$

$$\mathbf{b}_4 = \begin{bmatrix} 1 \; \beta_{2,1} \; \beta_{2,2} \; 0 \end{bmatrix}^T, \tag{4.5}$$

$\theta_{2,1}$ and $\theta_{2,2}$ are given by design, $0 < \theta_{2,1} \leq \pi$, $0 < \theta_{2,2} \leq \pi$, $\theta_{2,1} \neq \theta_{2,2}$, $0 \leq \beta_{2,1} \leq 1$, and $0 \leq \beta_{2,2} \leq 1$. Since $\mathbf{A}_4(\omega, \theta)$ is a full-rank square matrix, we deduce that the designed filter is

$$\mathbf{h}(\omega) = \mathbf{A}_4^{-1}(\omega, \theta)\, \mathbf{b}_4. \tag{4.6}$$

Once the second-order CDMA filter is derived with four microphones, it is easy to compute the beampattern:

$$|\mathcal{B}_2[\mathbf{h}(\omega), \theta]|^2 = \left|\mathbf{d}^H(\omega, \theta)\, \mathbf{A}_4^{-1}(\omega, \theta)\, \mathbf{b}_4\right|^2, \tag{4.7}$$

the directivity factor:

$$\mathcal{G}_{\mathrm{dn},2}[\mathbf{h}(\omega)] = \frac{1}{\mathbf{b}_4^T \mathbf{A}_4^{-H}(\omega, \theta)\, \mathbf{\Gamma}_{\mathrm{dn}}(\omega)\, \mathbf{A}_4^{-1}(\omega, \theta)\, \mathbf{b}_4}, \tag{4.8}$$

and the white noise gain:

$$\mathcal{G}_{\mathrm{wn},2}[\mathbf{h}(\omega)] = \frac{1}{\mathbf{b}_4^T \mathbf{A}_4^{-H}(\omega, \theta)\, \mathbf{A}_4^{-1}(\omega, \theta)\, \mathbf{b}_4}. \tag{4.9}$$

Now, we can numerically study the most well-known second-order CDMAs.

### 4.1.1 Second-Order Dipole

The dipole has a one at $\theta_s = 0$, a null at the angle $\theta_s + \pi/2$, and a one at the angle $\theta_s + \pi$. Hence, its parameters to solve (4.6) are $\theta_{2,1} = \pi/2$, $\theta_{2,2} = \pi$, $\beta_{2,1} = 0$, and $\beta_{2,2} = 1$.

Figure 4.1 displays the patterns of the second-order dipole with four microphones for several frequencies and two values of $\delta$ (1 and 2 cm). We can observe that they are identical to the desired second-order dipole shown in Fig. 2.3(a).

Figure 4.2 gives plots of $\mathcal{G}_{\mathrm{dn},2}[\mathbf{h}(\omega)]$ [where $\mathbf{h}(\omega)$ corresponds to the second-order dipole filter with four microphones], as a function of frequency, for different values of $\delta$. For small values of $\delta$, this directivity factor is constant and above 5 dB. For $\delta = 5$ cm, this directivity factor degrades pretty quickly above 2 kHz. It is of interest to observe that this gain is only slightly higher than the one obtained with the first-order dipole.

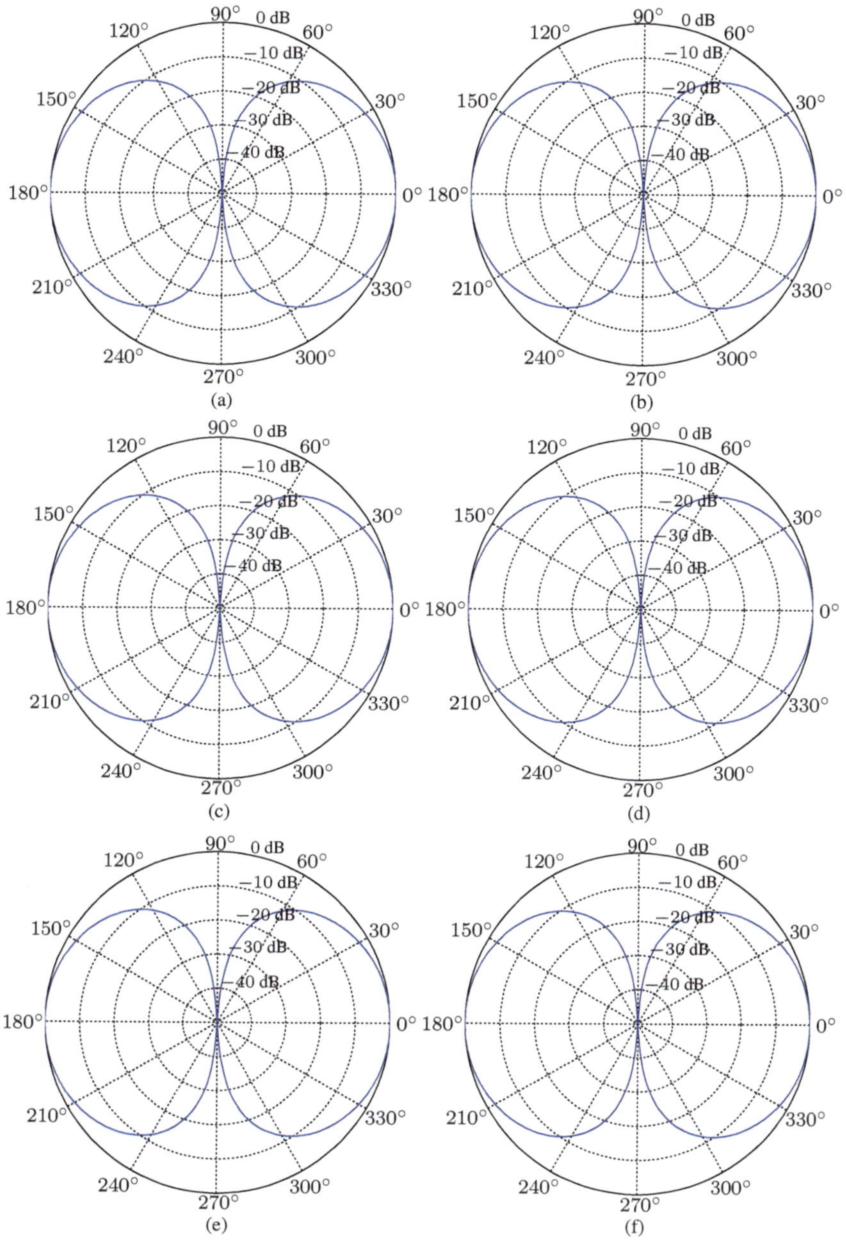


**Fig. 4.1** Patterns of the second-order dipole (with four microphones) for several frequencies and two values of $\delta$: (a) $f = 0.5$ kHz, $\delta = 1$ cm, (b) $f = 0.5$ kHz, $\delta = 2$ cm, (c) $f = 1$ kHz, $\delta = 1$ cm, (d) $f = 1$ kHz, $\delta = 2$ cm, (e) $f = 3$ kHz, $\delta = 1$ cm, and (f) $f = 3$ kHz, $\delta = 2$ cm.

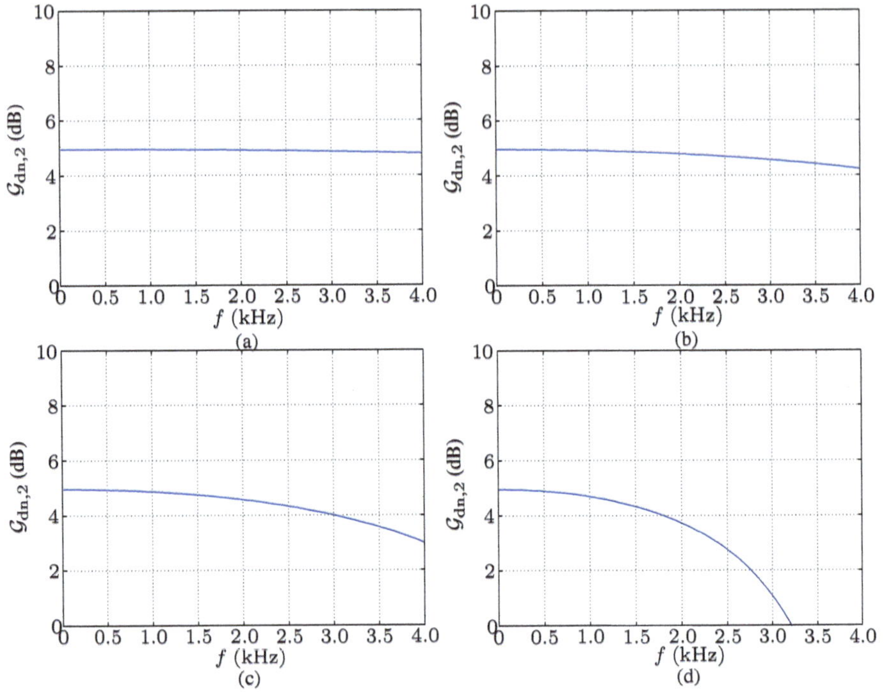


**Fig. 4.2** The directivity factor of the second-order dipole (with four microphones), as a function of frequency, for different values of $\delta$: (a) $\delta = 1$ cm, (b) $\delta = 2$ cm, (c) $\delta = 3$ cm, and (d) $\delta = 5$ cm.

Figure 4.3 gives plots of $\mathcal{G}_{wn,2}[\mathbf{h}(\omega)]$ [where $\mathbf{h}(\omega)$ corresponds to the second-order dipole filter with four microphones], as a function of frequency, for different values of $\delta$. As expected, this white noise gain is worse than the one corresponding to the first-order dipole.

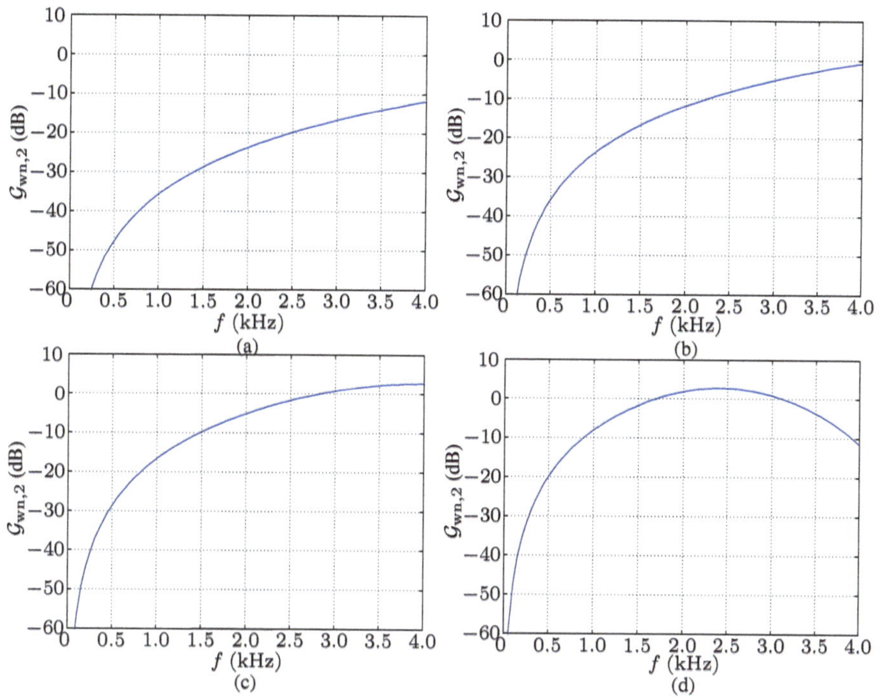


**Fig. 4.3** The white noise gain of the second-order dipole (with four microphones), as a function of frequency, for different values of $\delta$: (a) $\delta = 1$ cm, (b) $\delta = 2$ cm, (c) $\delta = 3$ cm, and (d) $\delta = 5$ cm.

We observe from simulations that a second-order dipole with a good compromise between the directivity factor and the white noise gain should be designed with an interelement spacing between 2 and 3 cm.

### 4.1.2 Second-Order Cardioid

In the second-order cardioid, there is a one at the angle $\theta_s = 0$ and two nulls at the angles $\theta_s + \pi/2$ and $\theta_s + \pi$. We see that $\theta_{2,1} = \pi/2$, $\theta_{2,2} = \pi$, $\beta_{2,1} = 0$, and $\beta_{2,2} = 0$. The filter corresponding to this cardioid is then easily obtained from (4.6).

Figure 4.4 shows the patterns of the second-order cardioid with four microphones for several frequencies and two values of $\delta$ (1 and 2 cm). The obtained patterns resemble the desired one shown in Fig. 2.3(b).

Figure 4.5 gives plots of $\mathcal{G}_{\mathrm{dn},2}[\mathbf{h}(\omega)]$ [where $\mathbf{h}(\omega)$ corresponds to the second-order cardioid filter with four microphones], as a function of frequency, for different values of $\delta$. This directivity factor is almost equal to 8 dB for

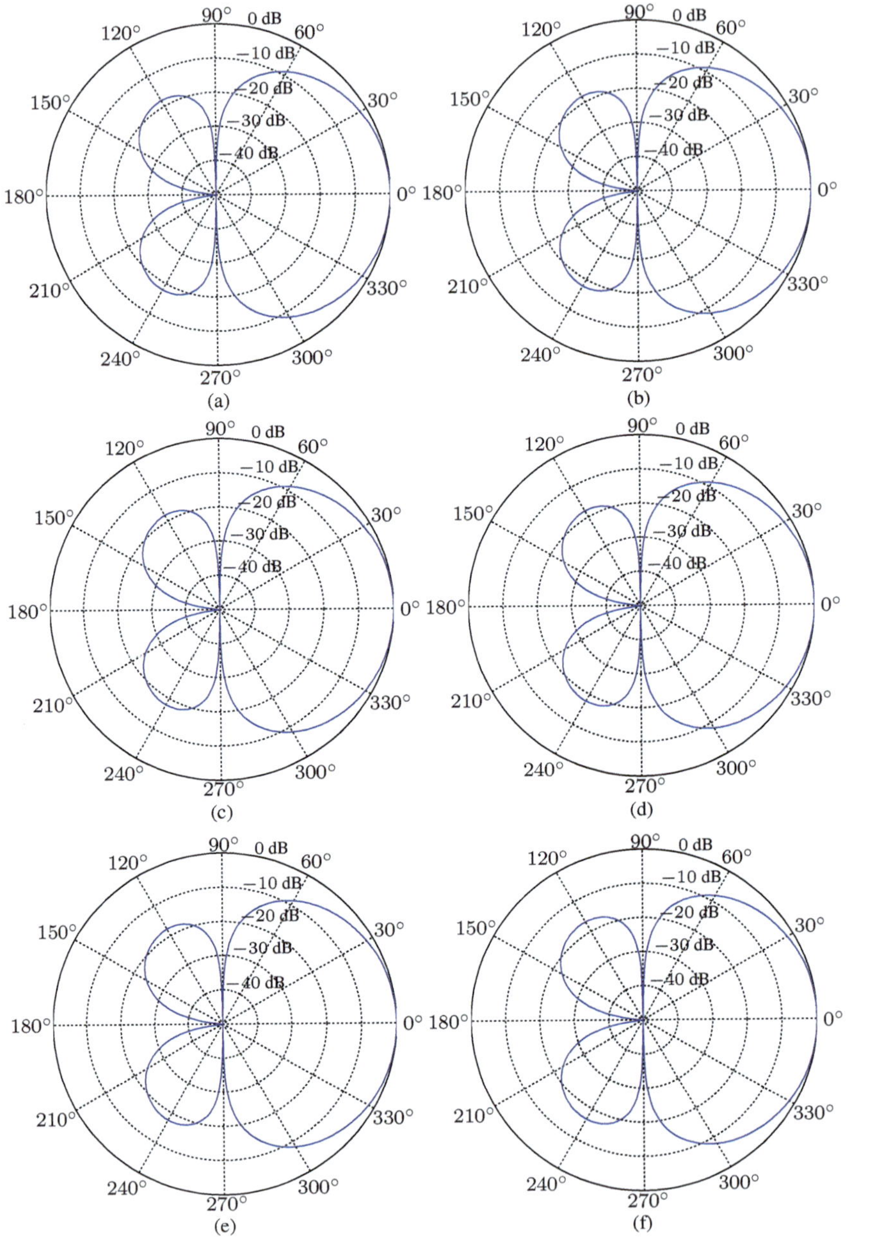

**Fig. 4.4** Patterns of the second-order cardioid (with four microphones) for several frequencies and two values of $\delta$: (a) $f = 0.5$ kHz, $\delta = 1$ cm, (b) $f = 0.5$ kHz, $\delta = 2$ cm, (c) $f = 1$ kHz, $\delta = 1$ cm, (d) $f = 1$ kHz, $\delta = 2$ cm, (e) $f = 3$ kHz, $\delta = 1$ cm, and (f) $f = 3$ kHz, $\delta = 2$ cm.

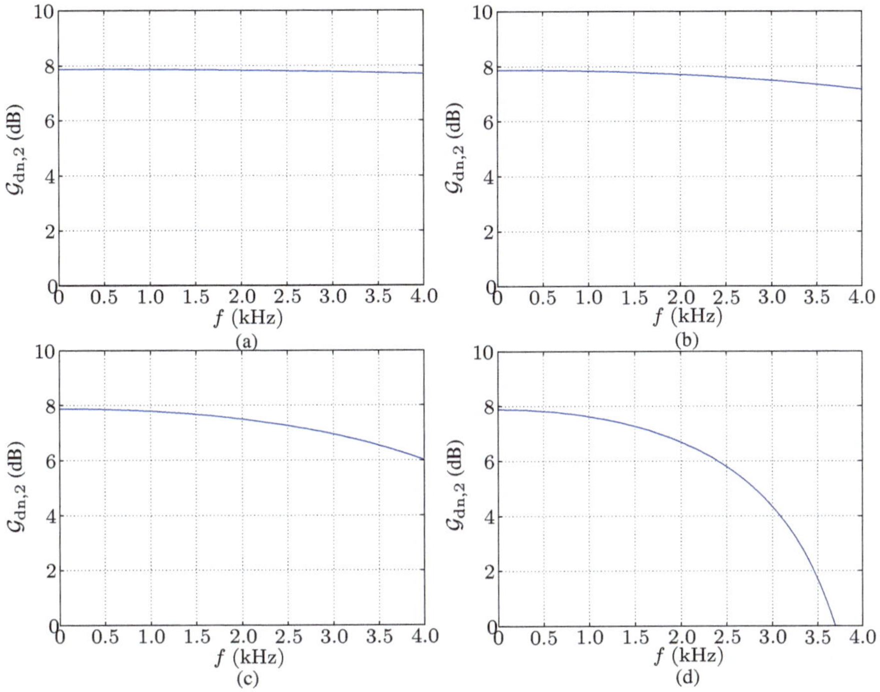


**Fig. 4.5** The directivity factor of the second-order cardioid (with four microphones), as a function of frequency, for different values of $\delta$: (a) $\delta = 1$ cm, (b) $\delta = 2$ cm, (c) $\delta = 3$ cm, and (d) $\delta = 5$ cm.

small values of $\delta$ and its value decreases, especially in high frequencies, when $\delta$ increases.

Figure 4.6 displays plots of $\mathcal{G}_{\mathrm{wn},2}[\mathbf{h}(\omega)]$ [where $\mathbf{h}(\omega)$ corresponds to the second-order cardioid filter with four microphones], as a function of frequency, for different values of $\delta$. This white noise gain is much better when $\delta$ increases.

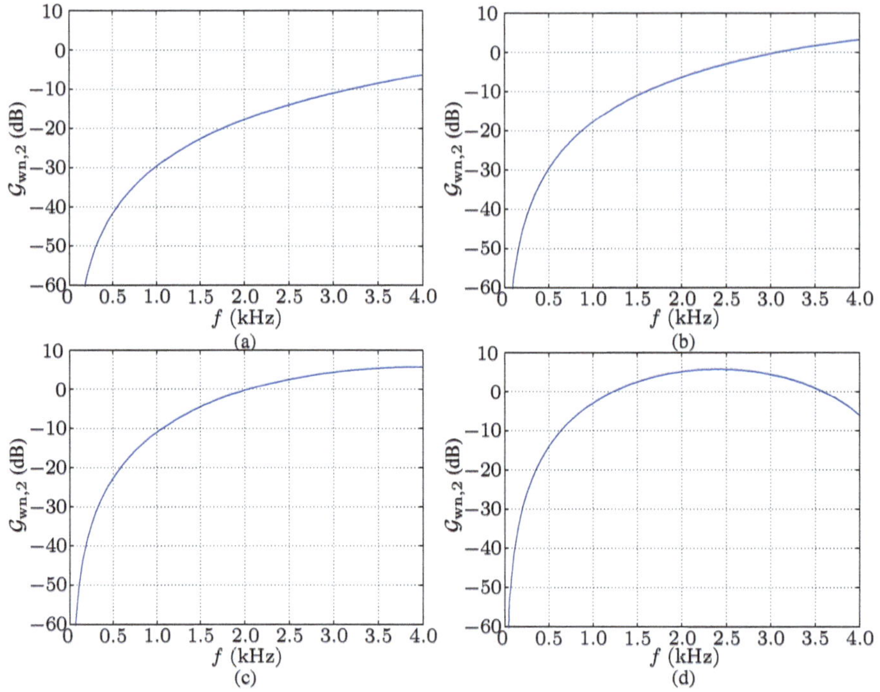


**Fig. 4.6** The white noise gain of the second-order cardioid (with four microphones), as a function of frequency, for different values of $\delta$: (a) $\delta = 1$ cm, (b) $\delta = 2$ cm, (c) $\delta = 3$ cm, and (d) $\delta = 5$ cm.

With a value of $\delta$ between 2 and 3 cm, we can design a reasonably good second-order cardioid with a 4-element UCA, which can steer in four different directions.

### *4.1.3 Second-Order Hypercardioid*

The second-order hypercardioid has a one at the angle $\theta_s = 0$ and two nulls at the angles $\theta_s + \pi/1.25$ and $\theta_s + \pi/2.5$. Therefore, we have $\theta_{2,1} = \pi/1.25$, $\theta_{2,2} = \pi/2.5$, $\beta_{2,1} = 0$, and $\beta_{2,2} = 0$; with these values, the corresponding filter follows immediately from (4.6).

Figure 4.7 shows the patterns of the second-order hypercardioid with four microphones for several frequencies and two values of $\delta$ (1 and 2 cm). The designed patterns are identical to the desired one given in Fig. 2.3(c).

Figure 4.8 gives plots of $\mathcal{G}_{\mathrm{dn},2}[\mathbf{h}(\omega)]$ [where $\mathbf{h}(\omega)$ corresponds to the second-order hypercardioid filter with four microphones], as a function of frequency, for different values of $\delta$. For small values of $\delta$, we can get a di-

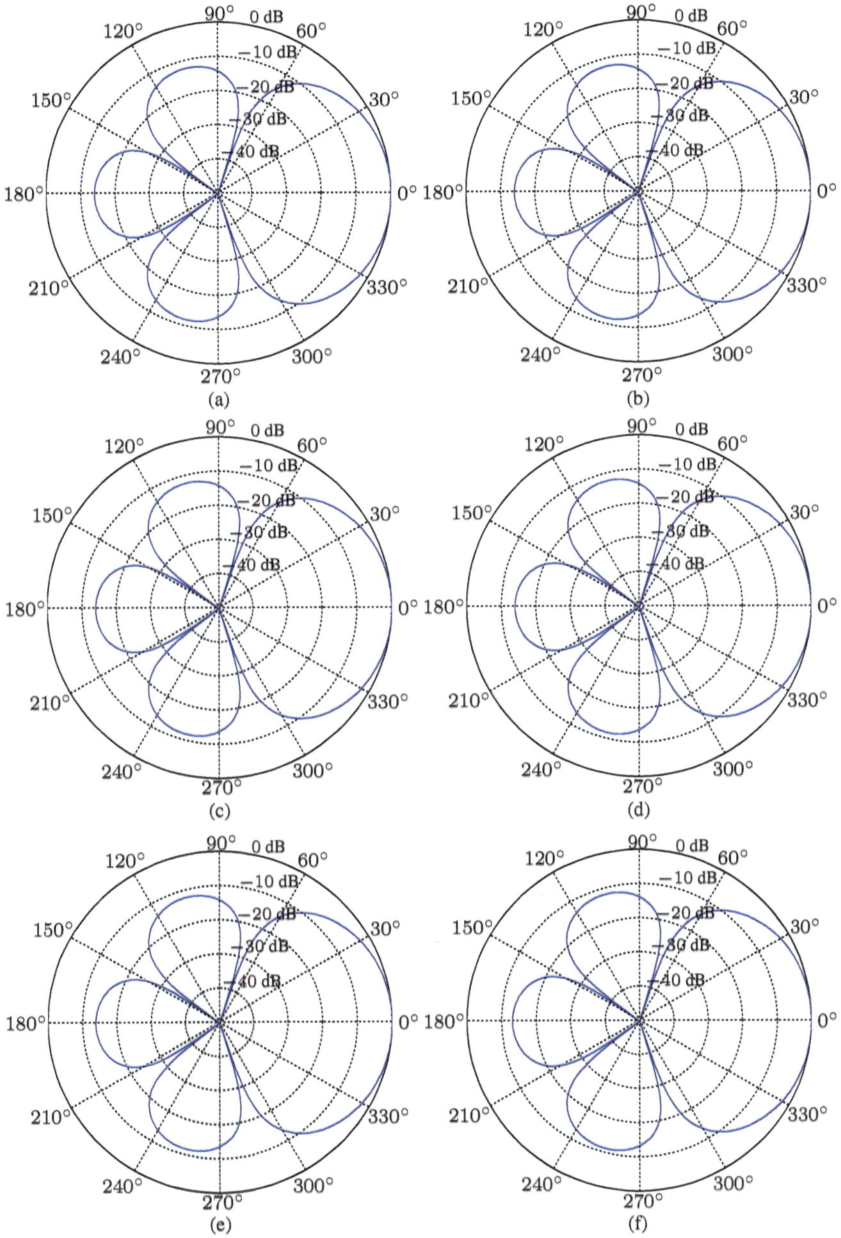


**Fig. 4.7** Patterns of the second-order hypercardioid (with four microphones) for several frequencies and two values of $\delta$: (a) $f = 0.5$ kHz, $\delta = 1$ cm, (b) $f = 0.5$ kHz, $\delta = 2$ cm, (c) $f = 1$ kHz, $\delta = 1$ cm, (d) $f = 1$ kHz, $\delta = 2$ cm, (e) $f = 3$ kHz, $\delta = 1$ cm, and (f) $f = 3$ kHz, $\delta = 2$ cm.

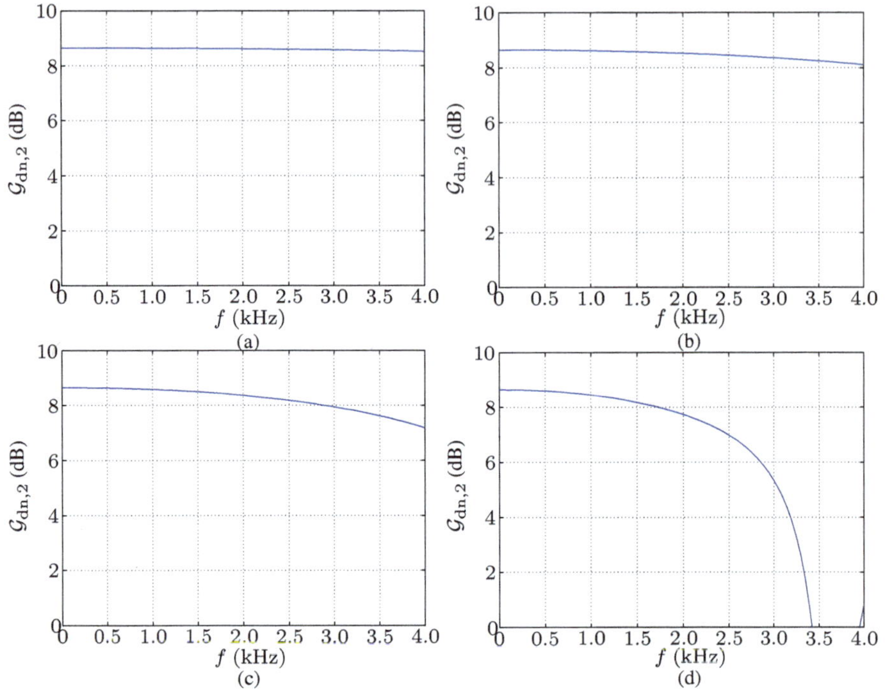


**Fig. 4.8** The directivity factor of the second-order hypercardioid (with four microphones), as a function of frequency, for different values of $\delta$: (a) $\delta = 1$ cm, (b) $\delta = 2$ cm, (c) $\delta = 3$ cm, and (d) $\delta = 5$ cm.

rectivity factor larger than 8 dB but for $\delta = 5$ cm, this directivity factor is negative in high frequencies.

Figure 4.9 displays plots of $\mathcal{G}_{\text{wn},2}\left[\mathbf{h}(\omega)\right]$ [where $\mathbf{h}(\omega)$ corresponds to the second-order hypercardioid filter with four microphones], as a function of frequency, for different values of $\delta$. This white noise gain behaves as expected.

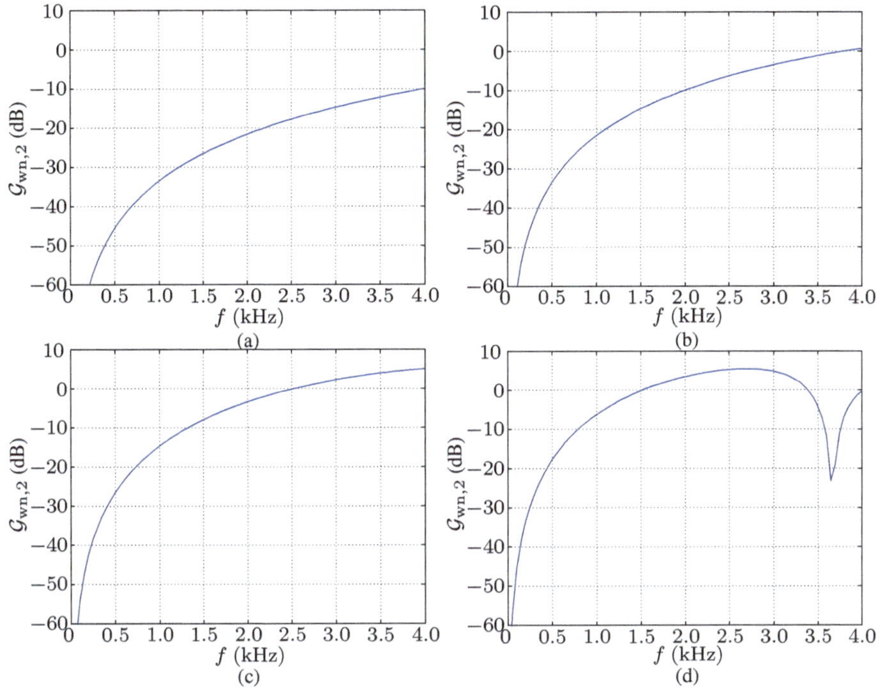


**Fig. 4.9** The white noise gain of the second-order hypercardioid (with four microphones), as a function of frequency, for different values of $\delta$: (a) $\delta = 1$ cm, (b) $\delta = 2$ cm, (c) $\delta = 3$ cm, and (d) $\delta = 5$ cm.

### *4.1.4 Second-Order Supercardioid*

The second-order supercardioid has a one at the angle $\theta_s = 0$ and two nulls at the angles $\theta_s + \pi/1.18$ and $\theta_s + \pi/1.69$. Therefore, we have $\theta_{2,1} = \pi/1.18$, $\theta_{2,2} = \pi/1.69$, $\beta_{2,1} = 0$, and $\beta_{2,2} = 0$; with these values, the corresponding filter follows immediately from (4.6).

Figure 4.10 shows the patterns of the second-order supercardioid with four microphones for several frequencies and two values of $\delta$ (1 and 2 cm). These patterns are the same as the ideal one [see Fig. 2.3(d)].

Figure 4.11 gives plots of $\mathcal{G}_{dn,2}[\mathbf{h}(\omega)]$ [where $\mathbf{h}(\omega)$ corresponds to the second-order supercardioid filter with four microphones], as a function of frequency, for different values of $\delta$. This directivity factor is above 7 dB but tends to deteriorate in high frequencies.

Figure 4.12 displays plots of $\mathcal{G}_{wn,2}[\mathbf{h}(\omega)]$ [where $\mathbf{h}(\omega)$ corresponds to the second-order supercardioid filter with four microphones], as a function of frequency, for different values of $\delta$. This white noise gain is smaller than 1 in low frequencies and for small values of $\delta$.

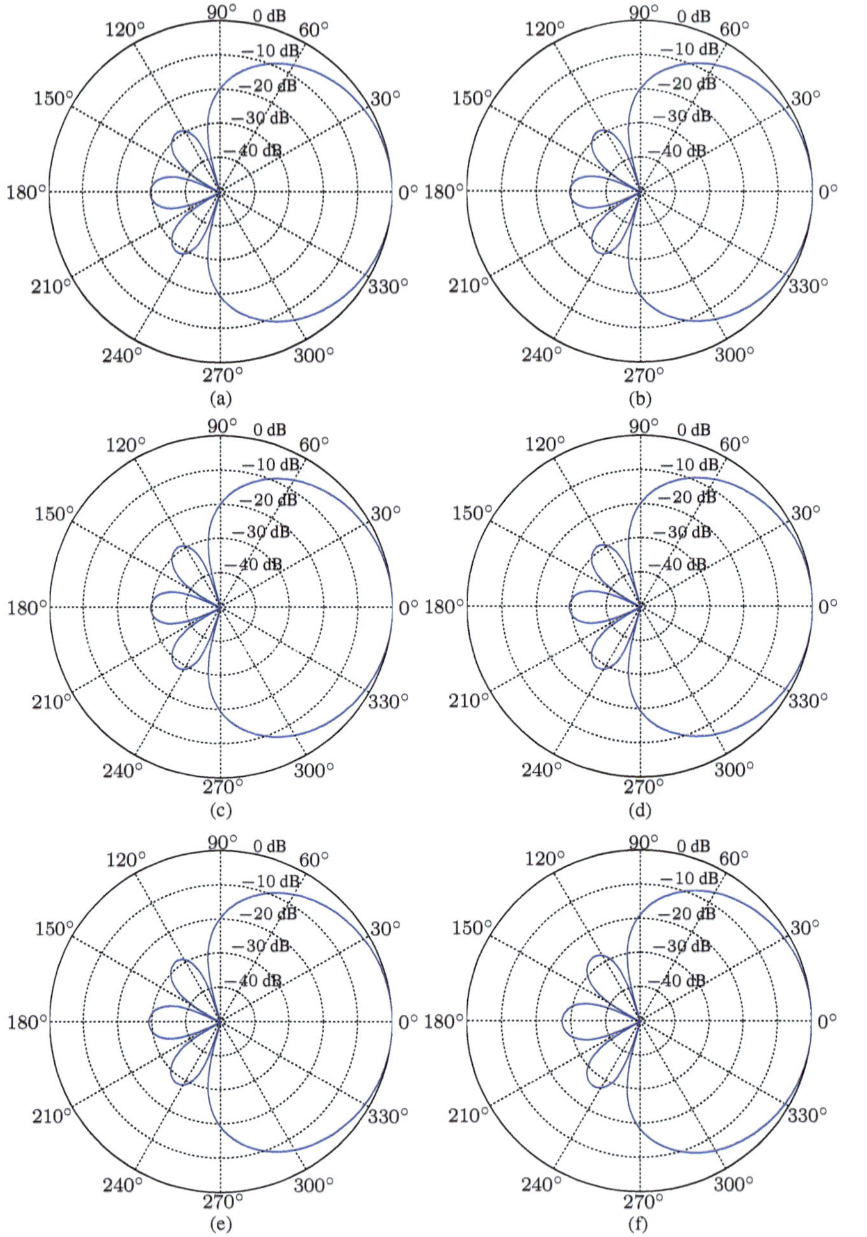


**Fig. 4.10** Patterns of the second-order supercardioid (with four microphones) for several frequencies and two values of $\delta$: (a) $f = 0.5$ kHz, $\delta = 1$ cm, (b) $f = 0.5$ kHz, $\delta = 2$ cm, (c) $f = 1$ kHz, $\delta = 1$ cm, (d) $f = 1$ kHz, $\delta = 2$ cm, (e) $f = 3$ kHz, $\delta = 1$ cm, and (f) $f = 3$ kHz, $\delta = 2$ cm.

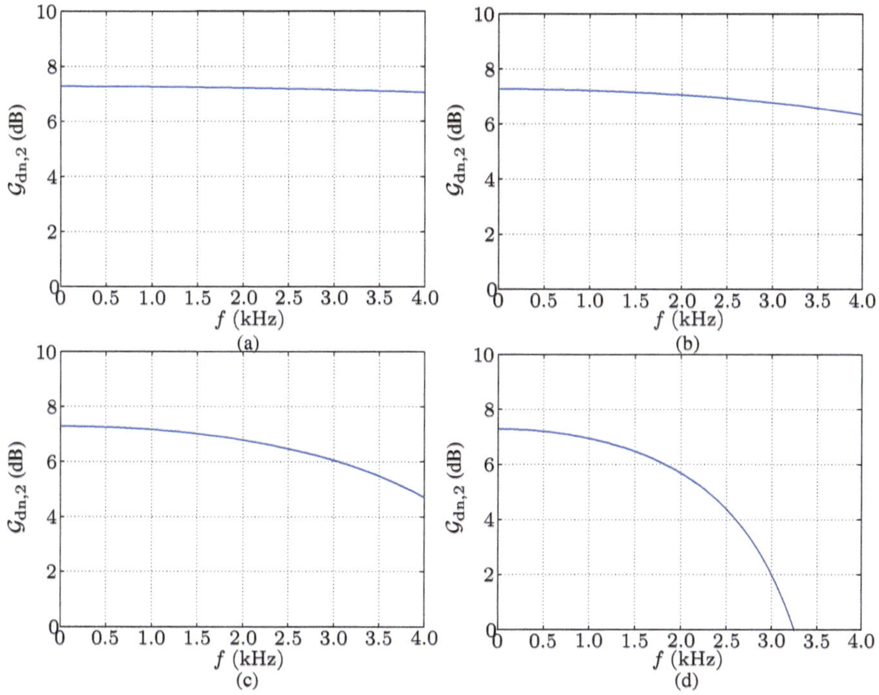


**Fig. 4.11** The directivity factor of the second-order supercardioid (with four microphones), as a function of frequency, for different values of $\delta$: (a) $\delta = 1$ cm, (b) $\delta = 2$ cm, (c) $\delta = 3$ cm, and (d) $\delta = 5$ cm.

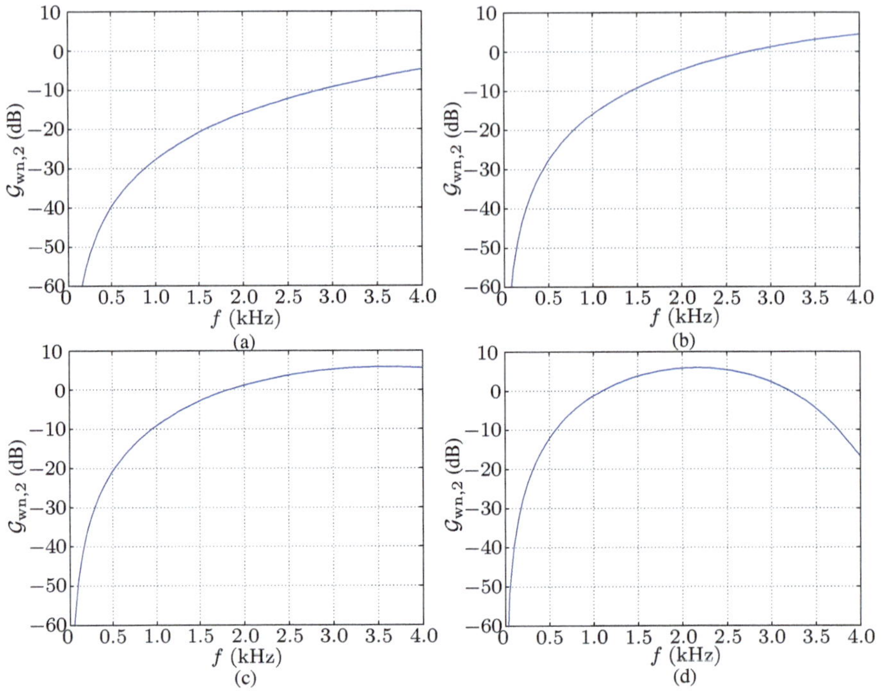


**Fig. 4.12** The white noise gain of the second-order supercardioid (with four microphones), as a function of frequency, for different values of $\delta$: (a) $\delta = 1$ cm, (b) $\delta = 2$ cm, (c) $\delta = 3$ cm, and (d) $\delta = 5$ cm.

Summarizing this section, we can definitely say that we can design any second-order directivity pattern with a 4-element UCA, which can steer in four different directions. Reasonable performance can be obtained in the range 0.5 to 4 kHz with an interelement spacing between 2 and 3 cm.

## 4.2 Principle and Design Examples with Five Microphones

The other obvious way to design a second-order CDMA is with $M = 5$. Now, the geometry of the array is a pentagon whose circumradius is $r$ and its sides length (or intersensor spacing) is $\delta = 2r\sin(\pi/5) = \frac{\sqrt{10-2\sqrt{5}}}{2}r$. We deduce that the positions of the five microphones are

$$\psi_1 = 0,\ \psi_2 = \frac{2\pi}{5},\ \psi_3 = \frac{4\pi}{5},\ \psi_4 = \frac{6\pi}{5},\ \psi_5 = \frac{8\pi}{5}, \tag{4.10}$$

and we can perfectly steer at the five different angles: $\theta_s = \psi_m$, $m = 1, 2, 3, 4, 5$. In all the rest, we choose $\theta_s = 0$. Because of the symmetry constraint, we must have $H_2(\omega) = H_5(\omega)$ and $H_3(\omega) = H_4(\omega)$. We deduce that the linear system of equations to solve is

$$\mathbf{A}_5(\omega,\theta)\,\mathbf{h}(\omega) = \mathbf{b}_5, \tag{4.11}$$

where

$$\mathbf{A}_5(\omega,\theta) = \begin{bmatrix} \mathbf{d}^H(\omega,0) \\ \mathbf{d}^H(\omega,\theta_{2,1}) \\ \mathbf{d}^H(\omega,\theta_{2,2}) \\ \mathbf{c}_{5,1}^T \\ \mathbf{c}_{5,2}^T \end{bmatrix}, \tag{4.12}$$

$$\mathbf{c}_{5,1} = \begin{bmatrix} 0 \; 1 \; 0 \; 0 \; -1 \end{bmatrix}^T, \tag{4.13}$$

$$\mathbf{c}_{5,2} = \begin{bmatrix} 0 \; 0 \; 1 \; -1 \; 0 \end{bmatrix}^T, \tag{4.14}$$

and

$$\mathbf{b}_5 = \begin{bmatrix} 1 \; \beta_{2,1} \; \beta_{2,2} \; 0 \; 0 \end{bmatrix}^T. \tag{4.15}$$

The parameters $\theta_{2,1}$, $\theta_{2,2}$, $\beta_{2,1}$, and $\beta_{2,2}$ are identical to the ones given in Section 4.1. Since $\mathbf{A}_5(\omega,\theta)$ is a full-rank square matrix, we deduce that the designed filter is

$$\mathbf{h}(\omega) = \mathbf{A}_5^{-1}(\omega,\theta)\,\mathbf{b}_5. \tag{4.16}$$

Once the second-order CDMA filter is derived with five microphones, it is easy to compute the beampattern, the directivity factor, and the white noise gain from (4.7), (4.8), and (4.9), respectively, by simply replacing $\mathbf{A}_4(\omega,\theta)$ with $\mathbf{A}_5(\omega,\theta)$, $\mathbf{b}_4$ with $\mathbf{b}_5$, and the steering vector of dimension four with its counterpart of dimension five.

Next, we study the second-order directivity patterns studied in the previous section.

### 4.2.1 Second-Order Dipole

The parameters of the second-order dipole with five microphones are identical to the ones given in Subsection 4.1.1 with four microphones.

Figure 4.13 displays the patterns of the second-order dipole with five microphones for several frequencies and two values of $\delta$ (1 and 2 cm). The designed patterns are identical to the ideal one [see Fig. 2.3(a)].

Figure 4.14 gives plots of $\mathcal{G}_{\mathrm{dn},2}[\mathbf{h}(\omega)]$ [where $\mathbf{h}(\omega)$ corresponds to the second-order dipole filter with five microphones], as a function of frequency, for different values of $\delta$. Comparing this directivity factor with the one obtained with four microphones, we can see that it is a bit worse, especially for large values of $\delta$.

Figure 4.15 gives plots of $\mathcal{G}_{\mathrm{wn},2}[\mathbf{h}(\omega)]$ [where $\mathbf{h}(\omega)$ corresponds to the second-order dipole filter with five microphones], as a function of frequency, for different values of $\delta$. The trend of this white noise gain is similar to the one built with four microphones.

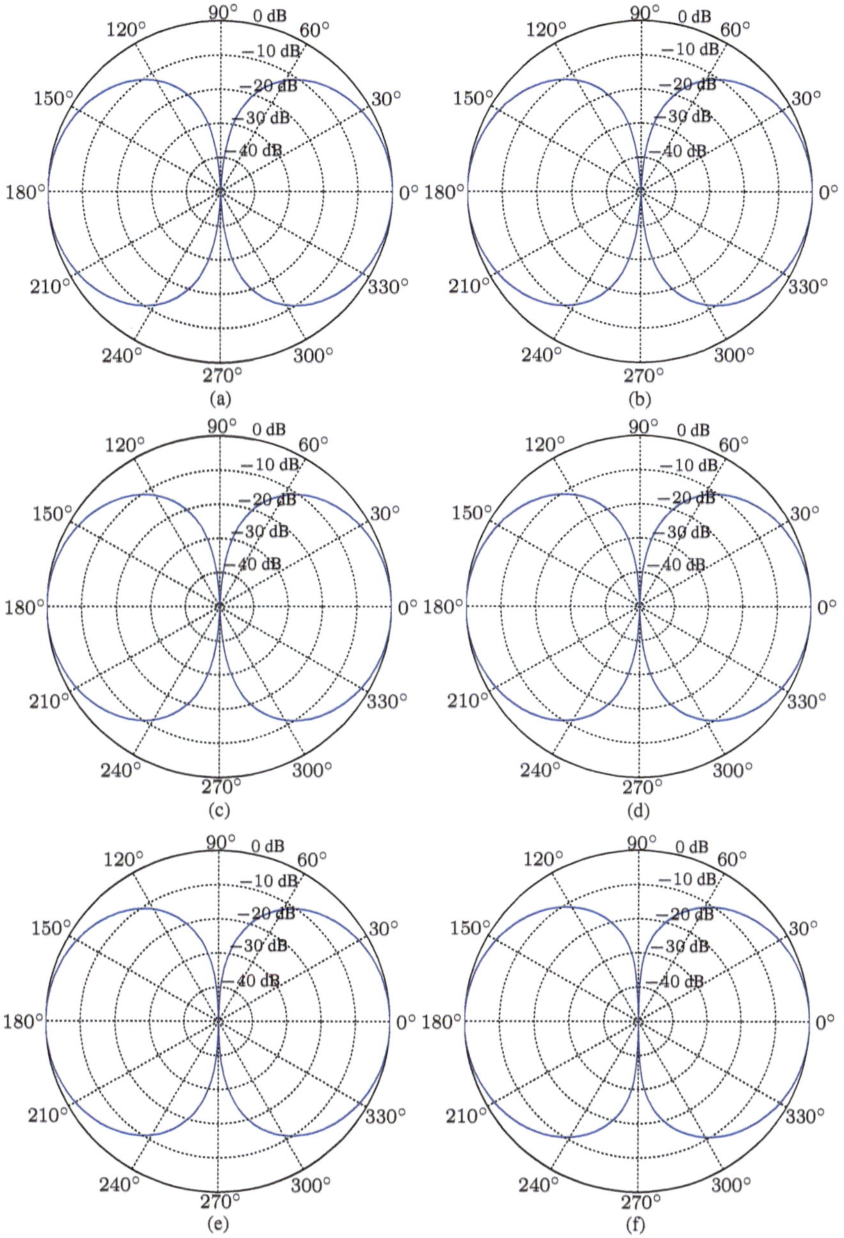


**Fig. 4.13** Patterns of the second-order dipole (with five microphones) for several frequencies and two values of $\delta$: (a) $f = 0.5$ kHz, $\delta = 1$ cm, (b) $f = 0.5$ kHz, $\delta = 2$ cm, (c) $f = 1$ kHz, $\delta = 1$ cm, (d) $f = 1$ kHz, $\delta = 2$ cm, (e) $f = 3$ kHz, $\delta = 1$ cm, and (f) $f = 3$ kHz, $\delta = 2$ cm.

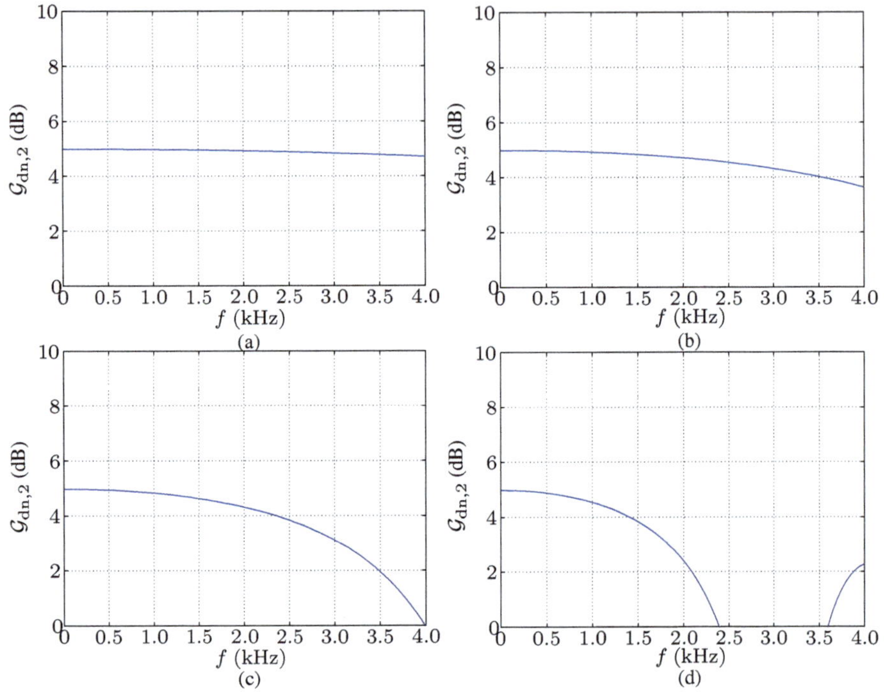


**Fig. 4.14** The directivity factor of the second-order dipole (with five microphones), as a function of frequency, for different values of $\delta$: (a) $\delta = 1$ cm, (b) $\delta = 2$ cm, (c) $\delta = 3$ cm, and (d) $\delta = 5$ cm.

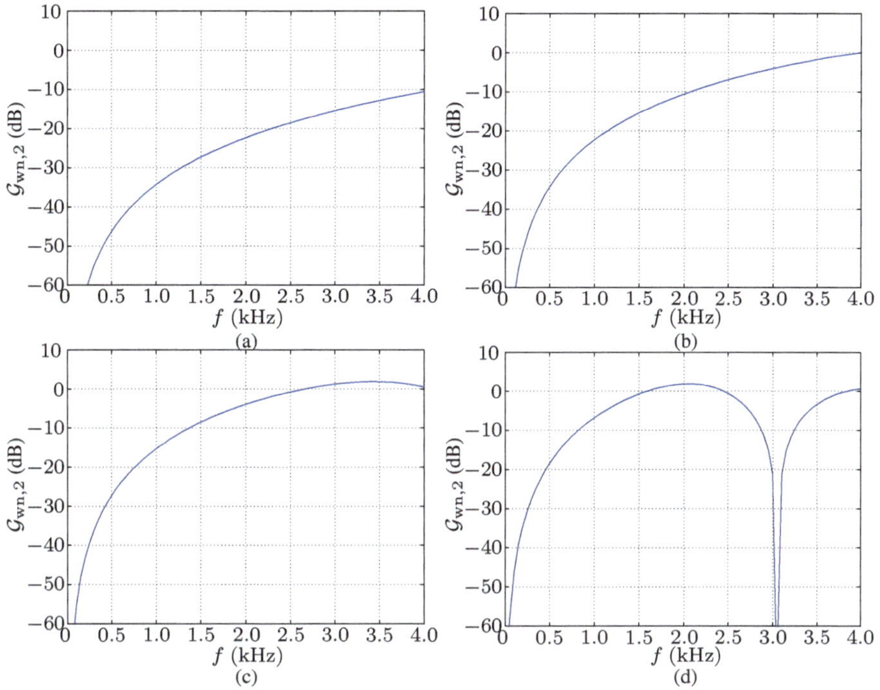


**Fig. 4.15** The white noise gain of the second-order dipole (with five microphones), as a function of frequency, for different values of $\delta$: (a) $\delta = 1$ cm, (b) $\delta = 2$ cm, (c) $\delta = 3$ cm, and (d) $\delta = 5$ cm.

## *4.2.2 Second-Order Cardioid*

The parameters of the second-order cardioid with five microphones are identical to the ones given in Subsection 4.1.2 with four microphones.

Figure 4.16 shows the patterns of the second-order cardioid with five microphones for several frequencies and two values of $\delta$ (1 and 2 cm). These patterns resemble the desired one shown in Fig. 2.3(b).

Figure 4.17 gives plots of $\mathcal{G}_{\mathrm{dn},2}[\mathbf{h}(\omega)]$ [where $\mathbf{h}(\omega)$ corresponds to the second-order cardioid filter with five microphones], as a function of frequency, for different values of $\delta$. This directivity factor is a little bit worse than the one designed with four microphones of Fig. 4.5.

Figure 4.18 displays plots of $\mathcal{G}_{\mathrm{wn},2}[\mathbf{h}(\omega)]$ [where $\mathbf{h}(\omega)$ corresponds to the second-order cardioid filter with five microphones], as a function of frequency, for different values of $\delta$. Overall, this white noise gain is similar to the one obtained with four microphones (see Fig. 4.6).

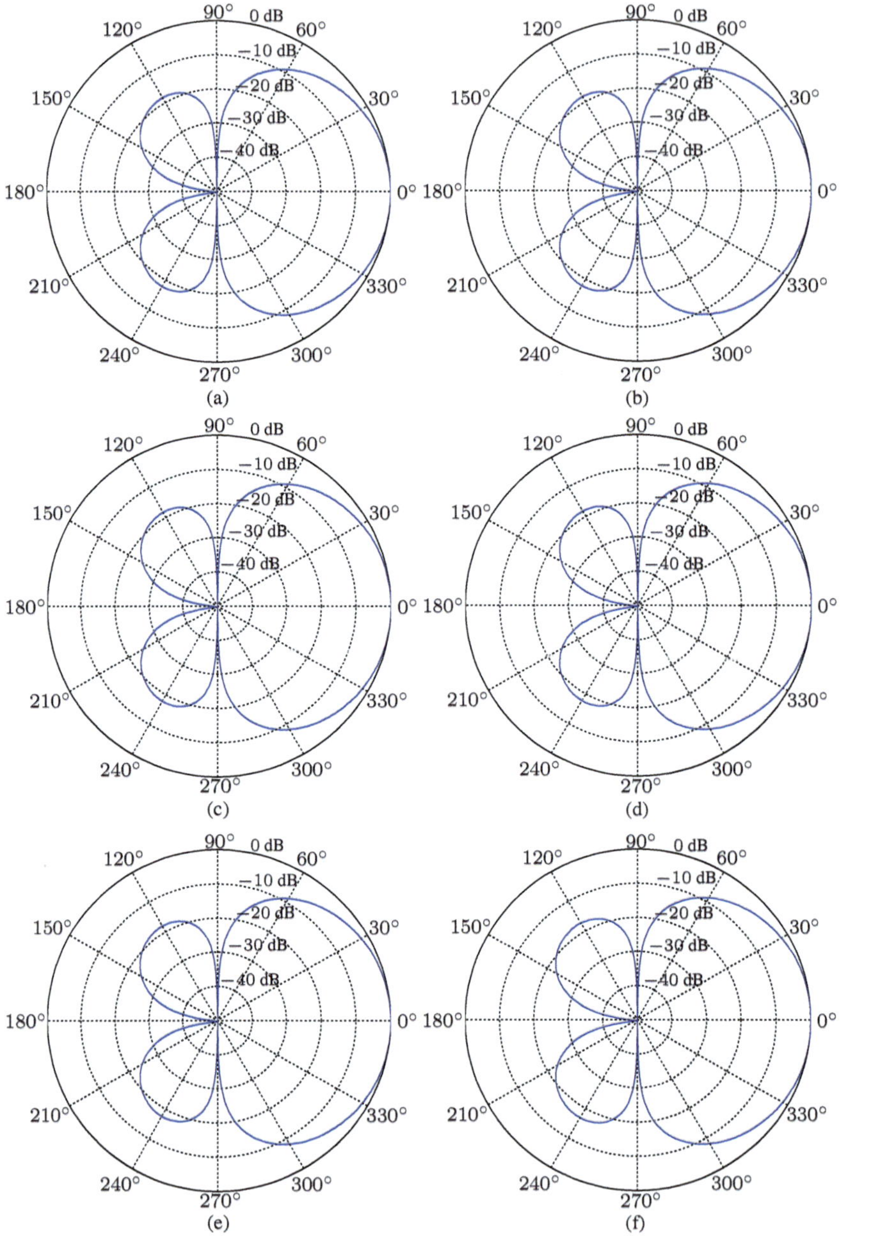


(a) (b) (c) (d) (e) (f)

**Fig. 4.16** Patterns of the second-order cardioid (with five microphones) for several frequencies and two values of $\delta$: (a) $f = 0.5$ kHz, $\delta = 1$ cm, (b) $f = 0.5$ kHz, $\delta = 2$ cm, (c) $f = 1$ kHz, $\delta = 1$ cm, (d) $f = 1$ kHz, $\delta = 2$ cm, (e) $f = 3$ kHz, $\delta = 1$ cm, and (f) $f = 3$ kHz, $\delta = 2$ cm.

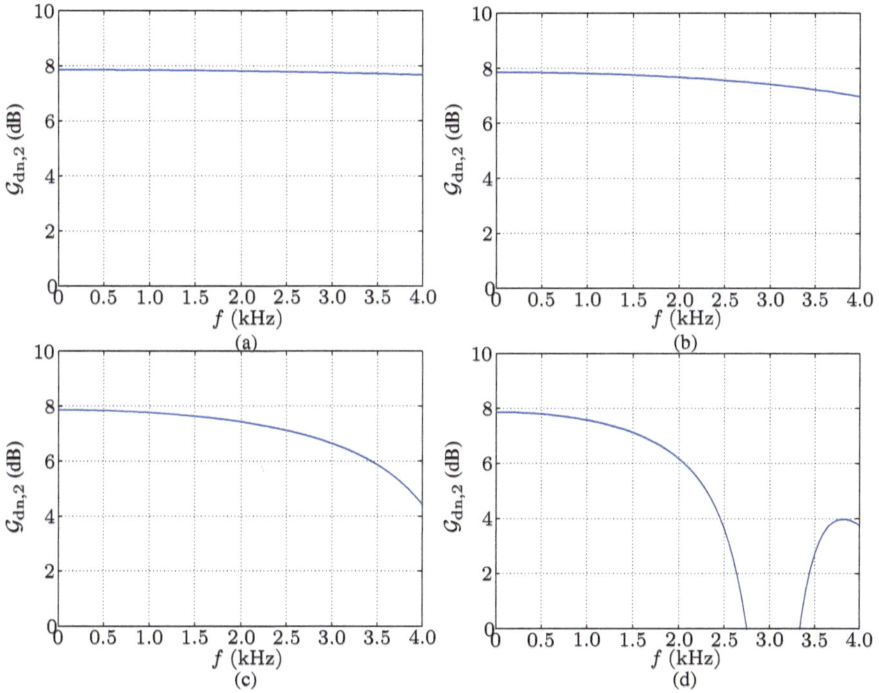


**Fig. 4.17** The directivity factor of the second-order cardioid (with five microphones), as a function of frequency, for different values of $\delta$: (a) $\delta = 1$ cm, (b) $\delta = 2$ cm, (c) $\delta = 3$ cm, and (d) $\delta = 5$ cm.

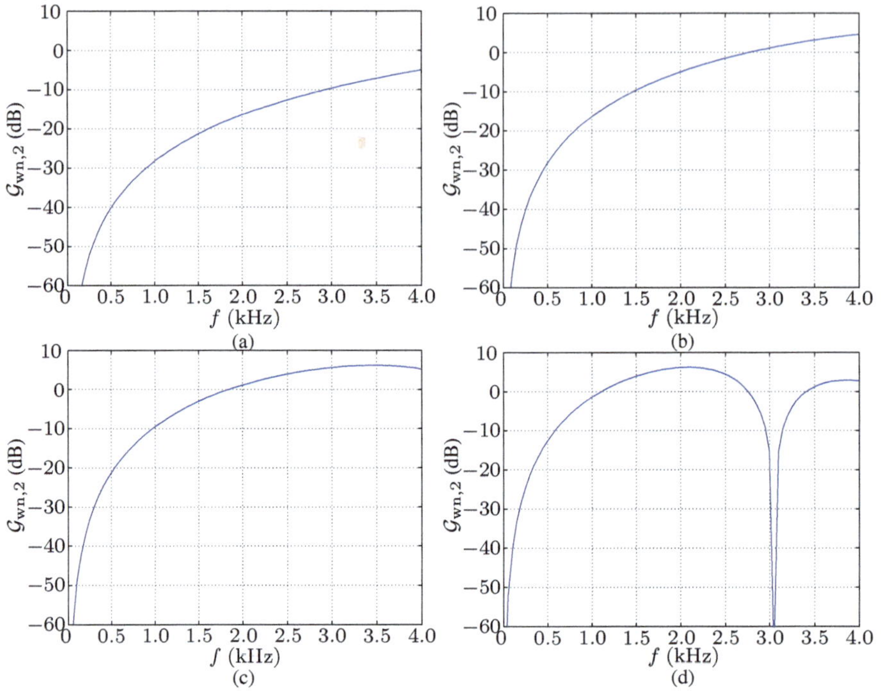


**Fig. 4.18** The white noise gain of the second-order cardioid (with five microphones), as a function of frequency, for different values of $\delta$: (a) $\delta = 1$ cm, (b) $\delta = 2$ cm, (c) $\delta = 3$ cm, and (d) $\delta = 5$ cm.

## 4.2.3 Second-Order Hypercardioid

The parameters of the second-order hypercardioid with five microphones are identical to the ones given in Subsection 4.1.3 with four microphones.

Figure 4.19 shows the patterns of the second-order hypercardioid with five microphones for several frequencies and two values of $\delta$ (1 and 2 cm). These hypercardioids are the same as the desired one given in Fig. 2.3(c).

Figure 4.20 gives plots of $\mathcal{G}_{\rm dn,2}\left[\mathbf{h}\left(\omega\right)\right]$ [where $\mathbf{h}\left(\omega\right)$ corresponds to the second-order hypercardioid filter with five microphones], as a function of frequency, for different values of $\delta$. This directivity factor is roughly the same as the one with four microphones.

Figure 4.21 displays plots of $\mathcal{G}_{\rm wn,2}\left[\mathbf{h}\left(\omega\right)\right]$ [where $\mathbf{h}\left(\omega\right)$ corresponds to the second-order hypercardioid filter with five microphones], as a function of frequency, for different values of $\delta$. We do not see much difference between this white noise gain and the one obtained with four microphones.

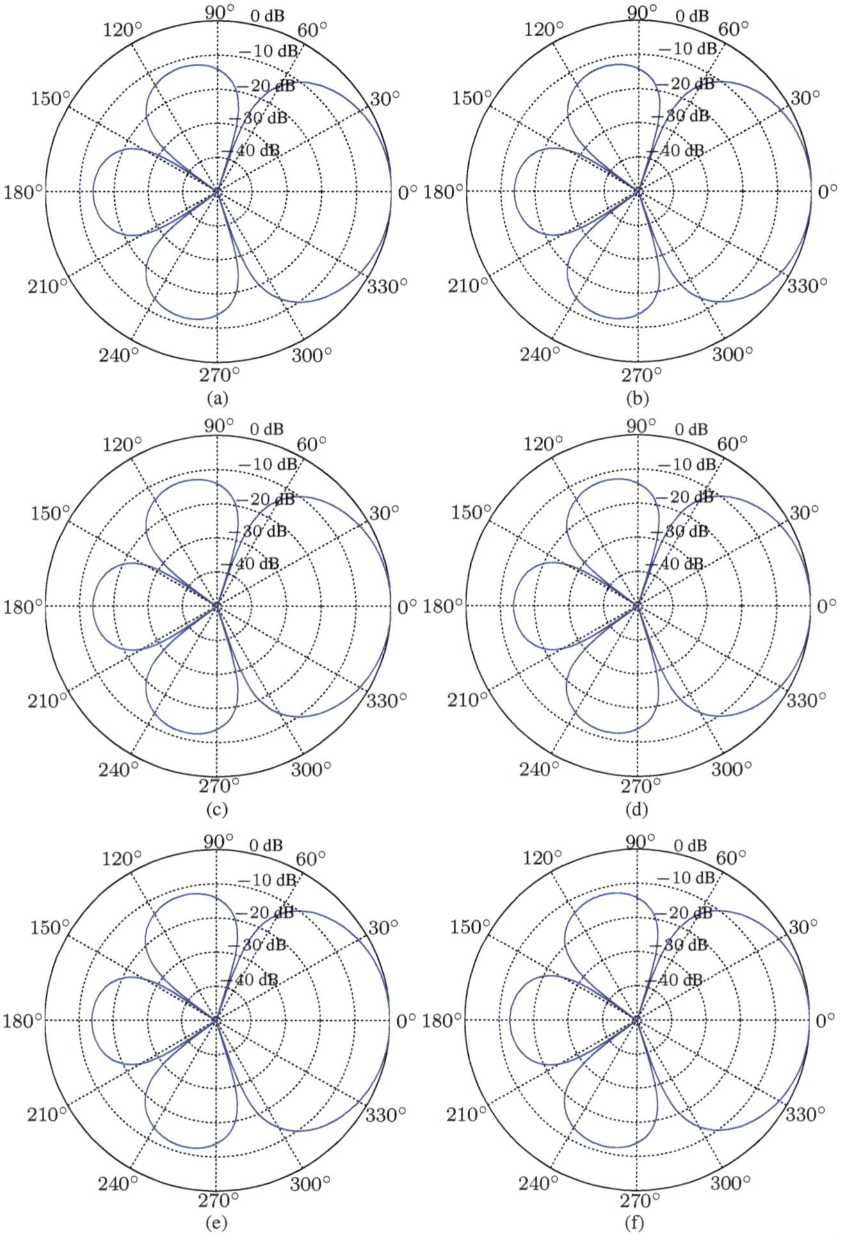


**Fig. 4.19** Patterns of the second-order hypercardioid (with five microphones) for several frequencies and two values of $\delta$: (a) $f = 0.5$ kHz, $\delta = 1$ cm, (b) $f = 0.5$ kHz, $\delta = 2$ cm, (c) $f = 1$ kHz, $\delta = 1$ cm, (d) $f = 1$ kHz, $\delta = 2$ cm, (e) $f = 3$ kHz, $\delta = 1$ cm, and (f) $f = 3$ kHz, $\delta = 2$ cm.

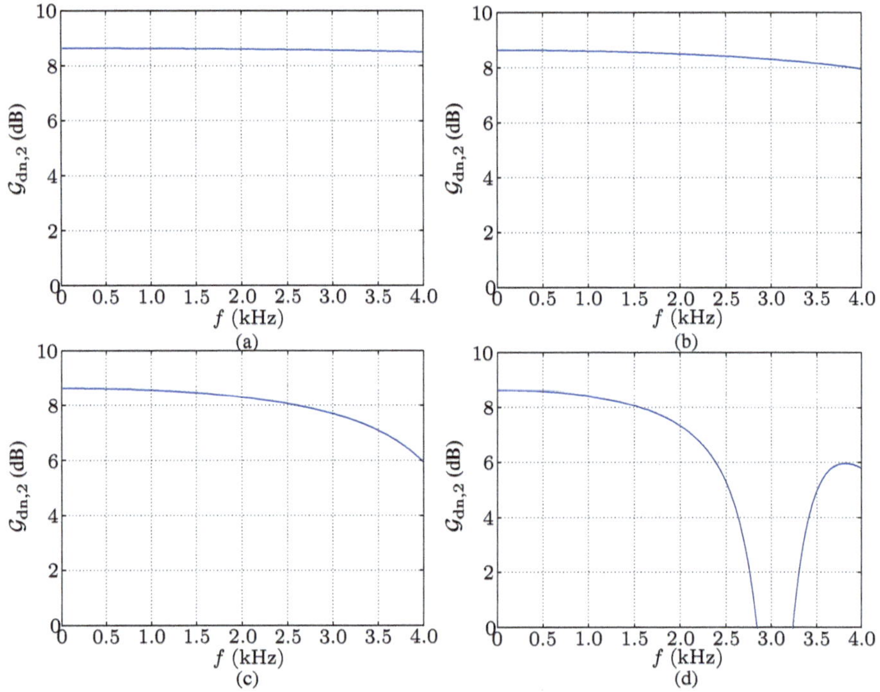


**Fig. 4.20** The directivity factor of the second-order hypercardioid (with five microphones), as a function of frequency, for different values of $\delta$: (a) $\delta = 1$ cm, (b) $\delta = 2$ cm, (c) $\delta = 3$ cm, and (d) $\delta = 5$ cm.

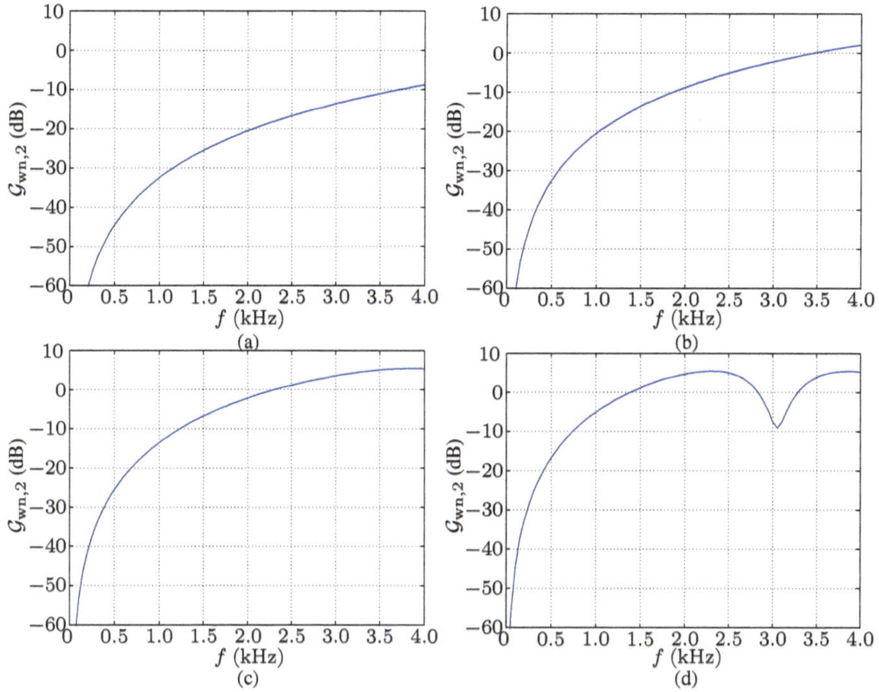


**Fig. 4.21** The white noise gain of the second-order hypercardioid (with five microphones), as a function of frequency, for different values of $\delta$: (a) $\delta = 1$ cm, (b) $\delta = 2$ cm, (c) $\delta = 3$ cm, and (d) $\delta = 5$ cm.

## *4.2.4 Second-Order Supercardioid*

The parameters of the second-order supercardioid with five microphones are identical to the ones given in Subsection 4.1.4 with four microphones.

Figure 4.22 shows the patterns of the second-order supercardioid with five microphones for several frequencies and two values of $\delta$ (1 and 2 cm). These patterns are identical to the ideal one [see Fig. 2.3(d)].

Figure 4.23 gives plots of $\mathcal{G}_{\text{dn},2}[\mathbf{h}(\omega)]$ [where $\mathbf{h}(\omega)$ corresponds to the second-order supercardioid filter with five microphones], as a function of frequency, for different values of $\delta$. This directivity factor is not much different from the one obtained with four microphones.

Figure 4.24 displays plots of $\mathcal{G}_{\text{wn},2}[\mathbf{h}(\omega)]$ [where $\mathbf{h}(\omega)$ corresponds to the second-order supercardioid filter with five microphones], as a function of frequency, for different values of $\delta$. This white noise gain is roughly the same as the one with four microphones.

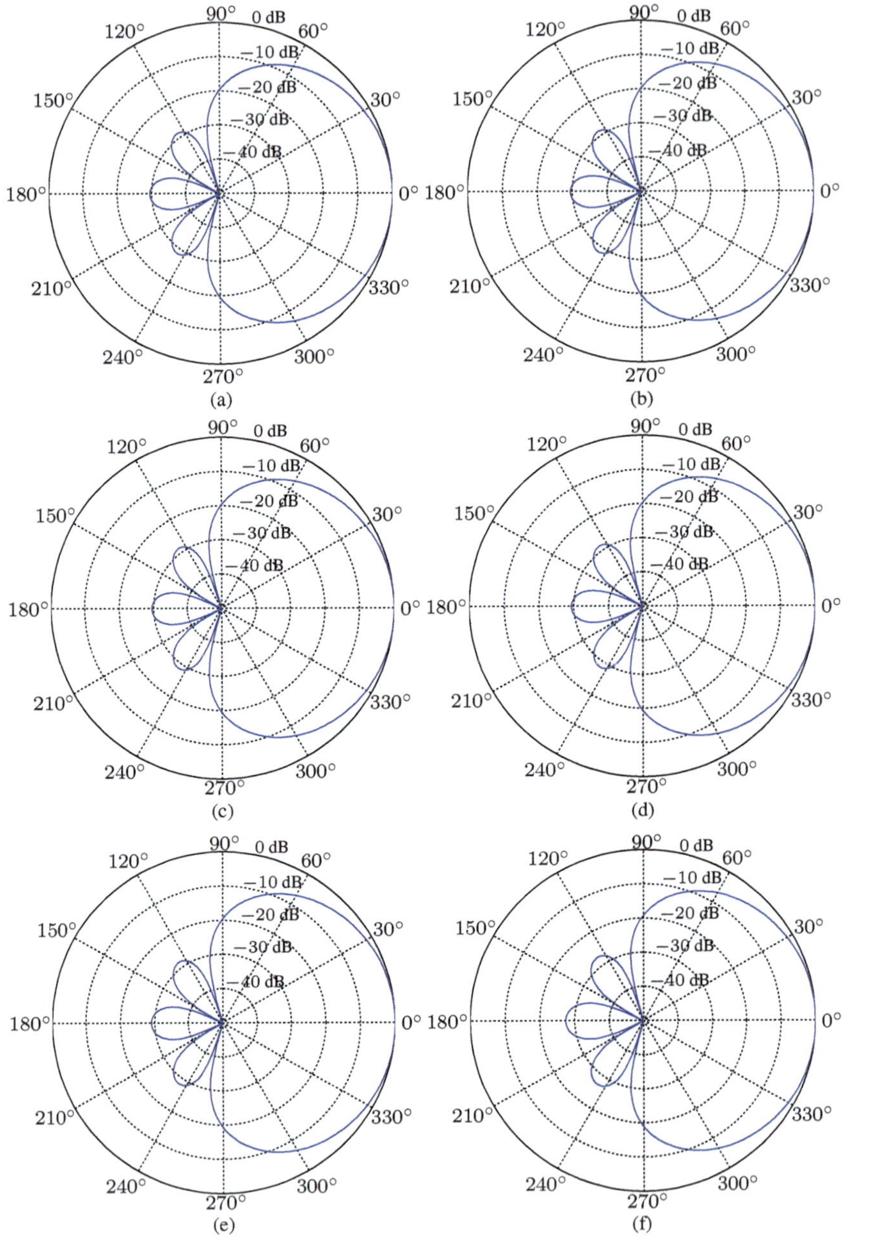


**Fig. 4.22** Patterns of the second-order supercardioid (with five microphones) for several frequencies and two values of $\delta$: (a) $f = 0.5$ kHz, $\delta = 1$ cm, (b) $f = 0.5$ kHz, $\delta = 2$ cm, (c) $f = 1$ kHz, $\delta = 1$ cm, (d) $f = 1$ kHz, $\delta = 2$ cm, (e) $f = 3$ kHz, $\delta = 1$ cm, and (f) $f = 3$ kHz, $\delta = 2$ cm.

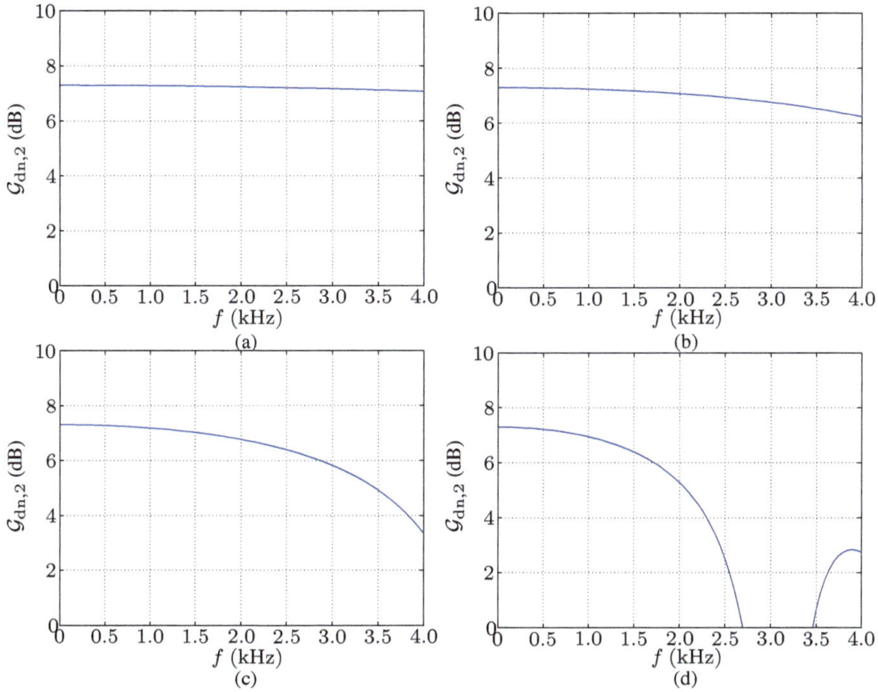


**Fig. 4.23** The directivity factor of the second-order supercardioid (with five microphones), as a function of frequency, for different values of $\delta$: (a) $\delta = 1$ cm, (b) $\delta = 2$ cm, (c) $\delta = 3$ cm, and (d) $\delta = 5$ cm.

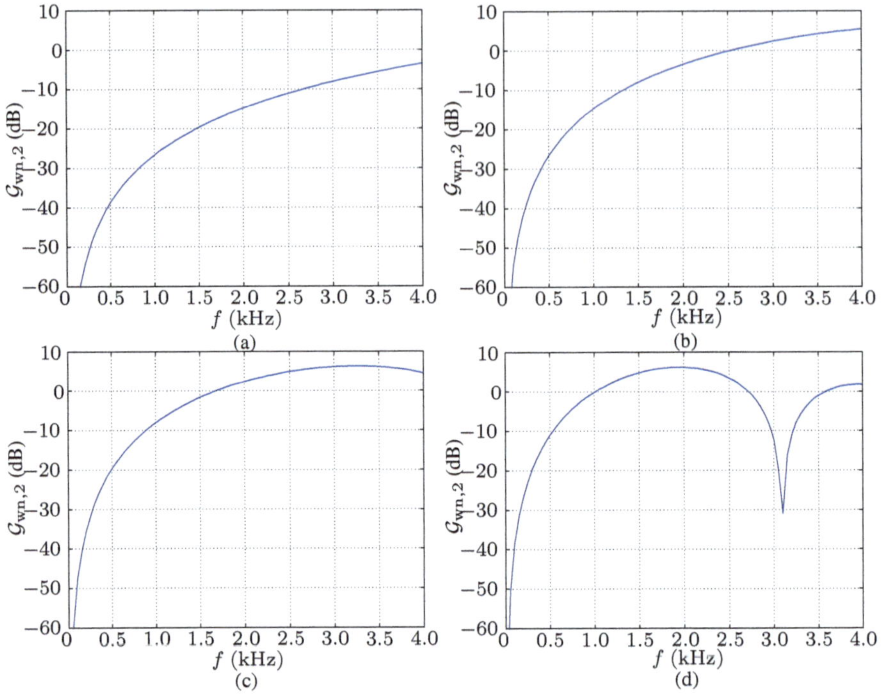


**Fig. 4.24** The white noise gain of the second-order supercardioid (with five microphones), as a function of frequency, for different values of $\delta$: (a) $\delta = 1$ cm, (b) $\delta = 2$ cm, (c) $\delta = 3$ cm, and (d) $\delta = 5$ cm.

To summarize this chapter, we can state that with a UCA we can design any second-order directivity pattern with four or five microphones. The performance with five microphones is slightly worse than the one with four microphones but, on the other hand, we can steer the array in one more direction.

## References


1. G. W. Elko, "Superdirectional microphone arrays," in *Acoustic Signal Processing for Telecommunication*, S. L. Gay and J. Benesty, Eds. Boston, MA: Kluwer Academic Publishers, 2000, Chapter 10, pp. 181–237.
2. J. Benesty and J. Chen, *Study and Design of Differential Microphone Arrays*. Berlin, Germany: Springer-Verlag, 2012.

# Chapter 5
# Design of Third-Order Circular Differential Arrays

In this chapter, we explore the design of third-order circular differential microphone arrays with six and seven sensors. In particular, we study a CDMA whose pattern has three distinct nulls [1]. We also generalize the approach to any CDMA order.

## 5.1 Principle and Design Example with Six Microphones

One way to design a third-order CDMA is with six microphones. In this case, the geometry of the array is a hexagon whose circumradius is $r$ and its sides length (or intersensor spacing) is $\delta = 2r\sin(\pi/6) = r$. We deduce that the positions of the six microphones are

$$\psi_1 = 0,\ \psi_2 = \frac{\pi}{3},\ \psi_3 = \frac{2\pi}{3},\ \psi_4 = \pi,\ \psi_5 = \frac{4\pi}{3},\ \psi_6 = \frac{5\pi}{3}, \tag{5.1}$$

and we can perfectly steer at the six different angles: $\theta_s = \psi_m$, $m = 1, 2, 3, 4, 5, 6$. In all the rest, we choose $\theta_s = 0$. Because of the symmetry constraint, we must have $H_2(\omega) = H_6(\omega)$ and $H_3(\omega) = H_5(\omega)$. Then, it is not hard to find that the linear system to solve is

$$\mathbf{A}_6(\omega, \theta)\,\mathbf{h}(\omega) = \mathbf{b}_6, \tag{5.2}$$

where

$$\mathbf{A}_6(\omega, \theta) = \begin{bmatrix} \mathbf{d}^H(\omega, 0) \\ \mathbf{d}^H(\omega, \theta_{3,1}) \\ \mathbf{d}^H(\omega, \theta_{3,2}) \\ \mathbf{d}^H(\omega, \theta_{3,3}) \\ \mathbf{c}_{6,1}^T \\ \mathbf{c}_{6,2}^T \end{bmatrix}, \tag{5.3}$$



J. Benesty et al., *Design of Circular Differential Microphone Arrays*,
Springer Topics in Signal Processing 12, DOI 10.1007/978-3-319-14842-7_5

$$\mathbf{c}_{6,1} = \begin{bmatrix} 0 \ 1 \ 0 \ 0 \ 0 \ -1 \end{bmatrix}^T, \tag{5.4}$$

$$\mathbf{c}_{6,2} = \begin{bmatrix} 0 \ 0 \ 1 \ 0 \ -1 \ 0 \end{bmatrix}^T, \tag{5.5}$$

$$\mathbf{b}_6 = \begin{bmatrix} 1 \ \beta_{3,1} \ \beta_{3,2} \ \beta_{3,3} \ 0 \ 0 \end{bmatrix}^T, \tag{5.6}$$

$\theta_{3,1}$, $\theta_{3,2}$, and $\theta_{3,3}$ are given by design, $0 < \theta_{3,1} \leq \pi$, $0 < \theta_{3,2} \leq \pi$, $0 < \theta_{3,3} \leq \pi$, $\theta_{3,1} \neq \theta_{3,2} \neq \theta_{3,3}$, $0 \leq \beta_{3,1} \leq 1$, $0 \leq \beta_{3,2} \leq 1$, and $0 \leq \beta_{3,3} \leq 1$. Since $\mathbf{A}_6(\omega, \theta)$ is a full-rank square matrix, we deduce that the designed filter is

$$\mathbf{h}(\omega) = \mathbf{A}_6^{-1}(\omega, \theta) \, \mathbf{b}_6. \tag{5.7}$$

Once the third-order CDMA filter is derived with six microphones, it is easy to compute the beampattern:

$$|\mathcal{B}_3[\mathbf{h}(\omega), \theta]|^2 = \left|\mathbf{d}^H(\omega, \theta) \, \mathbf{A}_6^{-1}(\omega, \theta) \, \mathbf{b}_6\right|^2, \tag{5.8}$$

the directivity factor:

$$\mathcal{G}_{\mathrm{dn},3}[\mathbf{h}(\omega)] = \frac{1}{\mathbf{b}_6^T \mathbf{A}_6^{-H}(\omega, \theta) \, \mathbf{\Gamma}_{\mathrm{dn}}(\omega) \, \mathbf{A}_6^{-1}(\omega, \theta) \, \mathbf{b}_6}, \tag{5.9}$$

and the white noise gain:

$$\mathcal{G}_{\mathrm{wn},3}[\mathbf{h}(\omega)] = \frac{1}{\mathbf{b}_6^T \mathbf{A}_6^{-H}(\omega, \theta) \, \mathbf{A}_6^{-1}(\omega, \theta) \, \mathbf{b}_6}. \tag{5.10}$$

We are now going to numerically study a CDMA whose beampattern has three distinct nulls. As usual, there is a one at $\theta_s$ and we take the three distinct nulls at the angles $\theta_s + \pi/2$, $\theta_s + 2\pi/3$, and $\theta_s + \pi$. Hence, the parameters to solve (5.7) are $\theta_{3,1} = \pi/2$, $\theta_{3,2} = 2\pi/3$, $\theta_{3,3} = \pi$, and $\beta_{3,1} = \beta_{3,2} = \beta_{3,3} = 0$.

Figure 5.1 displays the patterns of the third-order CDMA with six microphones for several frequencies and two values of $\delta$ (1 and 2 cm). The designed patterns are as expected and they all look the same.

Figure 5.2 gives plots of $\mathcal{G}_{\mathrm{dn},3}[\mathbf{h}(\omega)]$ [where $\mathbf{h}(\omega)$ corresponds to the third-order CDMA filter with six microphones], as a function of frequency, for different values of $\delta$. For comparison, this directivity factor, for small values of $\delta$, is only slightly higher than the directivity factor of the second-order supercardioid.

Figure 5.3 gives plots of $\mathcal{G}_{\mathrm{wn},3}[\mathbf{h}(\omega)]$ [where $\mathbf{h}(\omega)$ corresponds to the third-order CDMA filter with six microphones], as a function of frequency, for different values of $\delta$. For comparison, this white noise gain is significantly worse than the white noise gain of the second-order supercardioid.

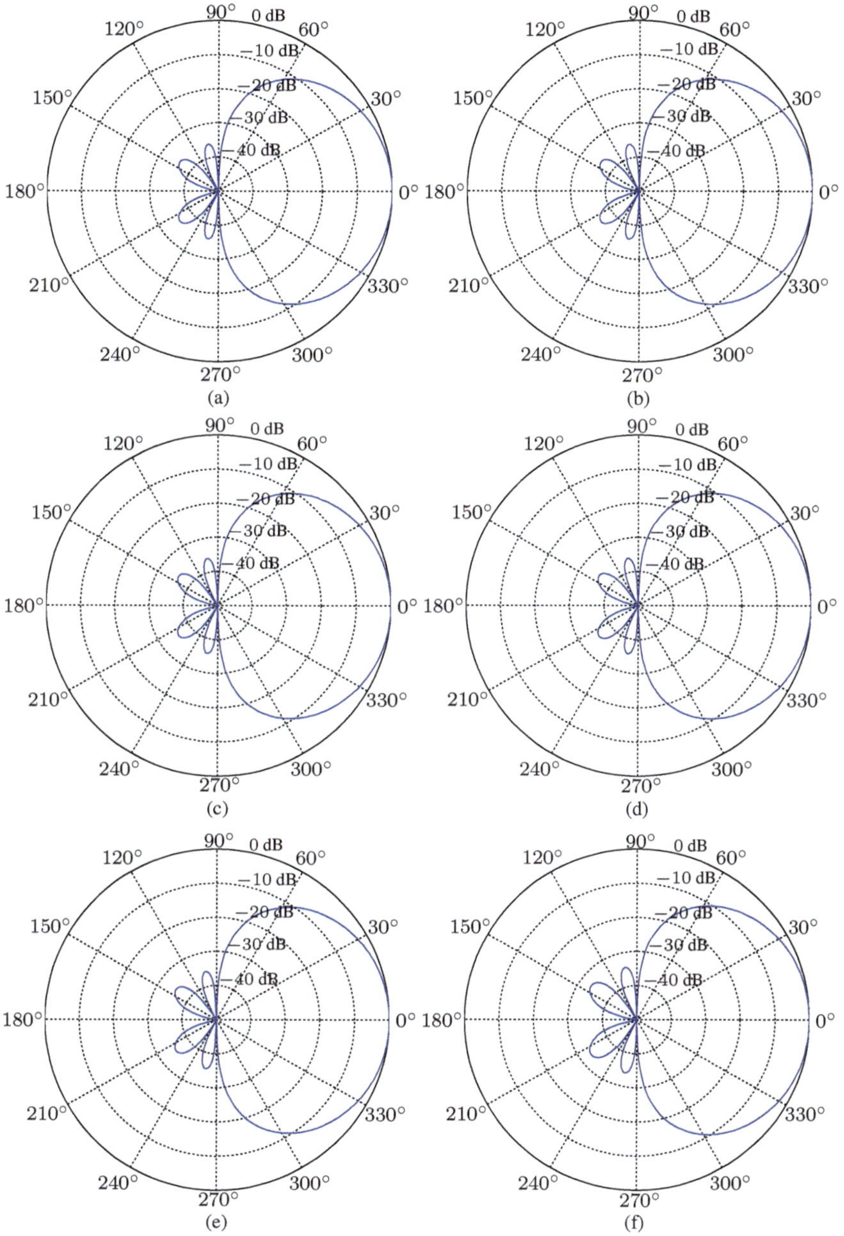


**Fig. 5.1** Patterns of the third-order CDMA (with six microphones) for several frequencies and two values of $\delta$: (a) $f = 0.5$ kHz, $\delta = 1$ cm, (b) $f = 0.5$ kHz, $\delta = 2$ cm, (c) $f = 1$ kHz, $\delta = 1$ cm, (d) $f = 1$ kHz, $\delta = 2$ cm, (e) $f = 3$ kHz, $\delta = 1$ cm, and (f) $f = 3$ kHz, $\delta = 2$ cm.

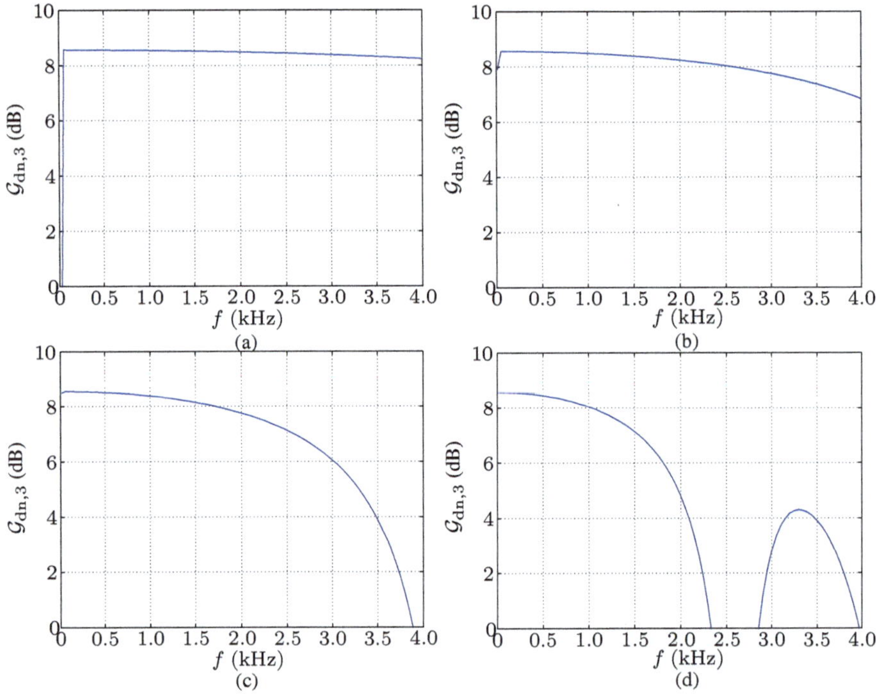


**Fig. 5.2** The directivity factor of the third-order CDMA (with six microphones), as a function of frequency, for different values of $\delta$: (a) $\delta = 1$ cm, (b) $\delta = 2$ cm, (c) $\delta = 3$ cm, and (d) $\delta = 5$ cm.

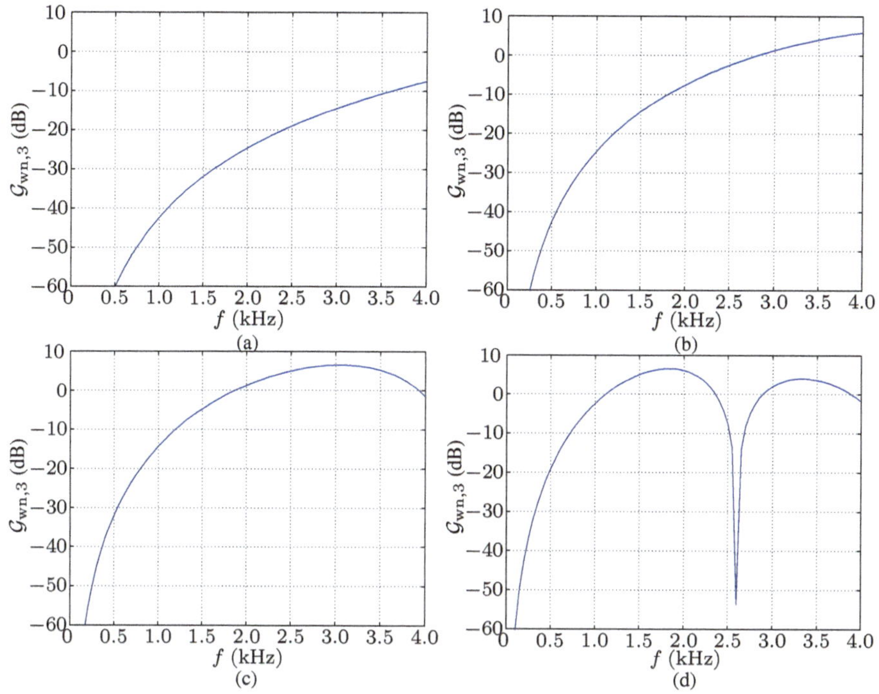


**Fig. 5.3** The white noise gain of the third-order CDMA (with six microphones), as a function of frequency, for different values of $\delta$: (a) $\delta = 1$ cm, (b) $\delta = 2$ cm, (c) $\delta = 3$ cm, and (d) $\delta = 5$ cm.

## 5.2 Principle and Design Example with Seven Microphones

The other obvious way to design a third-order CDMA is with $M = 7$. Now, the geometry of the array is a heptagon whose circumradius is $r$ and its sides length (or intersensor spacing) is $\delta = 2r \sin(\pi/7) \approx 0.868r$. We deduce that the positions of the seven microphones are

$$\psi_1 = 0,\ \psi_2 = \frac{2\pi}{7},\ \psi_3 = \frac{4\pi}{7},\ \psi_4 = \frac{6\pi}{7},\ \psi_5 = \frac{8\pi}{7},\ \psi_6 = \frac{10\pi}{7},\ \psi_7 = \frac{12\pi}{7}, \tag{5.11}$$

and we can perfectly steer at the seven different angles: $\theta_{\mathrm{s}} = \psi_m$, $m = 1, 2, 3, 4, 5, 6, 7$. In all the rest, we choose $\theta_{\mathrm{s}} = 0$. Because of the symmetry constraint, we must have $H_2(\omega) = H_7(\omega)$, $H_3(\omega) = H_6(\omega)$, and $H_4(\omega) = H_5(\omega)$. We deduce that the linear system of equations to solve is

$$\mathbf{A}_7(\omega, \theta)\, \mathbf{h}(\omega) = \mathbf{b}_7, \tag{5.12}$$

where

$$\mathbf{A}_7(\omega,\theta)=\begin{bmatrix}\mathbf{d}^H(\omega,0)\\ \mathbf{d}^H(\omega,\theta_{3,1})\\ \mathbf{d}^H(\omega,\theta_{3,2})\\ \mathbf{d}^H(\omega,\theta_{3,3})\\ \mathbf{c}_{7,1}^T\\ \mathbf{c}_{7,2}^T\\ \mathbf{c}_{7,3}^T\end{bmatrix}, \tag{5.13}$$

$$\mathbf{c}_{7,1}=\begin{bmatrix}0\ 1\ 0\ 0\ 0\ 0\ -1\end{bmatrix}^T, \tag{5.14}$$

$$\mathbf{c}_{7,2}=\begin{bmatrix}0\ 0\ 1\ 0\ 0\ -1\ 0\end{bmatrix}^T, \tag{5.15}$$

$$\mathbf{c}_{7,3}=\begin{bmatrix}0\ 0\ 0\ 1\ -1\ 0\ 0\end{bmatrix}^T, \tag{5.16}$$

and

$$\mathbf{b}_7=\begin{bmatrix}1\ \beta_{3,1}\ \beta_{3,2}\ \beta_{3,3}\ 0\ 0\ 0\end{bmatrix}^T. \tag{5.17}$$

The parameters $\theta_{3,1}$, $\theta_{3,2}$, $\theta_{3,3}$, $\beta_{3,1}$, $\beta_{3,2}$, and $\beta_{3,3}$ are identical to the ones given in Section 5.1. Since $\mathbf{A}_7(\omega,\theta)$ is a full-rank square matrix, we deduce that the designed filter is

$$\mathbf{h}(\omega)=\mathbf{A}_7^{-1}(\omega,\theta)\,\mathbf{b}_7. \tag{5.18}$$

Once the second-order CDMA filter is derived with seven microphones, it is easy to compute the beampattern, the directivity factor, and the white noise gain from (5.8), (5.9), and (5.10), respectively, by simply replacing $\mathbf{A}_6(\omega,\theta)$ with $\mathbf{A}_7(\omega,\theta)$, $\mathbf{b}_6$ with $\mathbf{b}_7$, and the steering vector of dimension six with its counterpart of dimension seven.

We study again the example with three distinct nulls given in Section 5.1. Figure 5.4 displays the patterns of the third-order CDMA with seven microphones for several frequencies and two values of $\delta$ (1 and 2 cm). These patterns are identical to the ones obtained with six microphones.

Figure 5.5 gives plots of $\mathcal{G}_{\mathrm{dn},3}[\mathbf{h}(\omega)]$ [where $\mathbf{h}(\omega)$ corresponds to the third-order CDMA filter with seven microphones], as a function of frequency, for different values of $\delta$. This directivity factor is slightly worse than the one obtained with six microphones (see Fig. 5.2).

Figure 5.6 gives plots of $\mathcal{G}_{\mathrm{wn},3}[\mathbf{h}(\omega)]$ [where $\mathbf{h}(\omega)$ corresponds to the third-order CDMA filter with seven microphones], as a function of frequency, for different values of $\delta$. There is not significant differences between this white noise gain and the one obtained with six microphones (see Fig. 5.3).

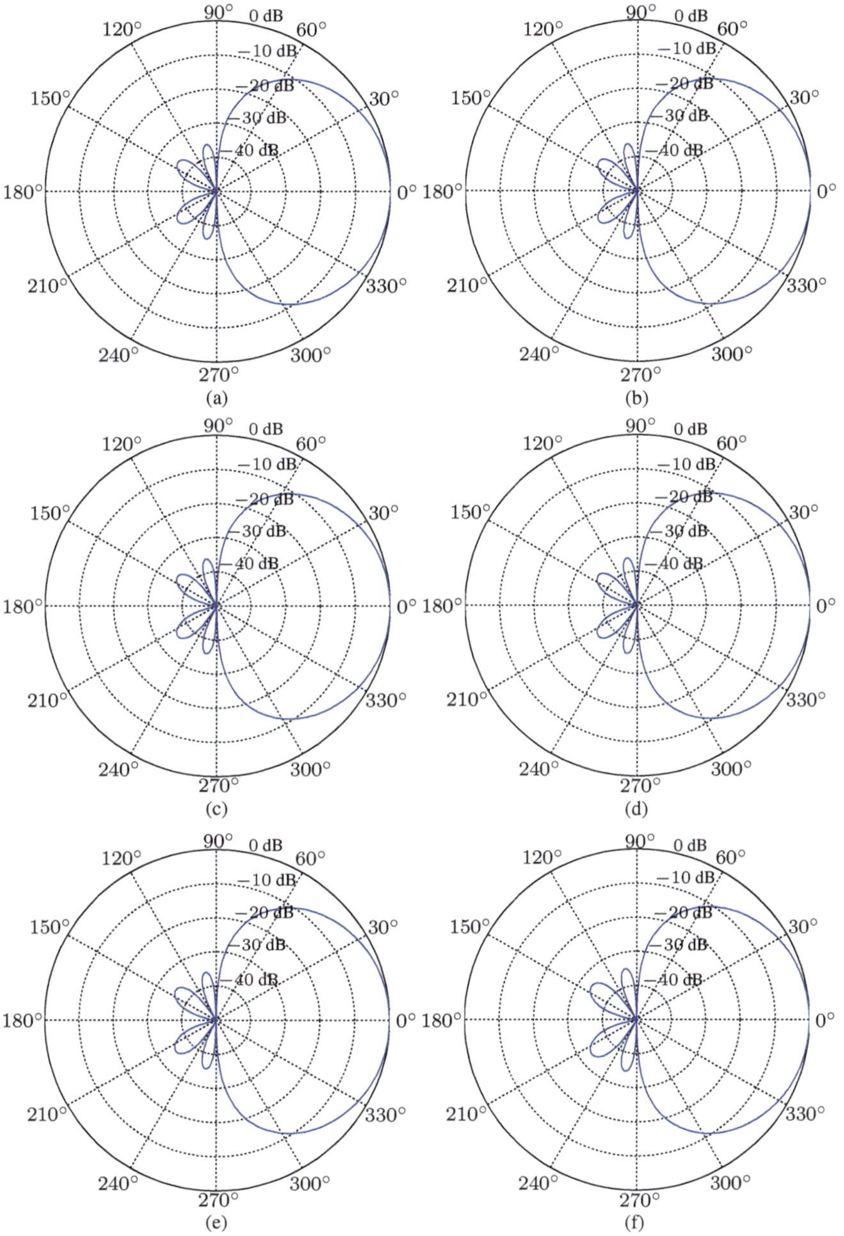


**Fig. 5.4** Patterns of the third-order CDMA (with seven microphones) for several frequencies and two values of $\delta$: (a) $f = 0.5$ kHz, $\delta = 1$ cm, (b) $f = 0.5$ kHz, $\delta = 2$ cm, (c) $f = 1$ kHz, $\delta = 1$ cm, (d) $f = 1$ kHz, $\delta = 2$ cm, (e) $f = 3$ kHz, $\delta = 1$ cm, and (f) $f = 3$ kHz, $\delta = 2$ cm.

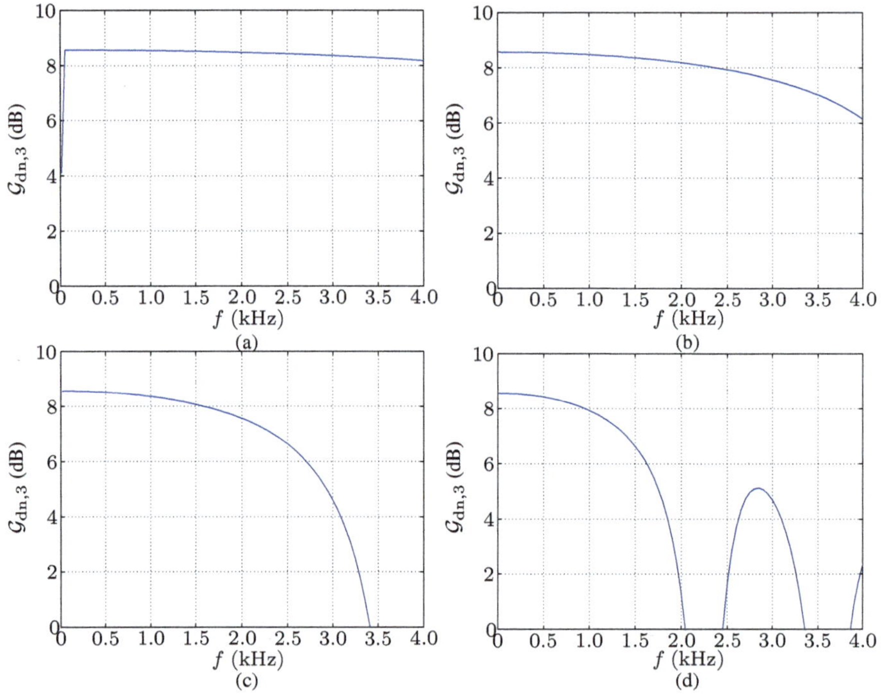


**Fig. 5.5** The directivity factor of the third-order CDMA (with seven microphones), as a function of frequency, for different values of $\delta$: (a) $\delta = 1$ cm, (b) $\delta = 2$ cm, (c) $\delta = 3$ cm, and (d) $\delta = 5$ cm.

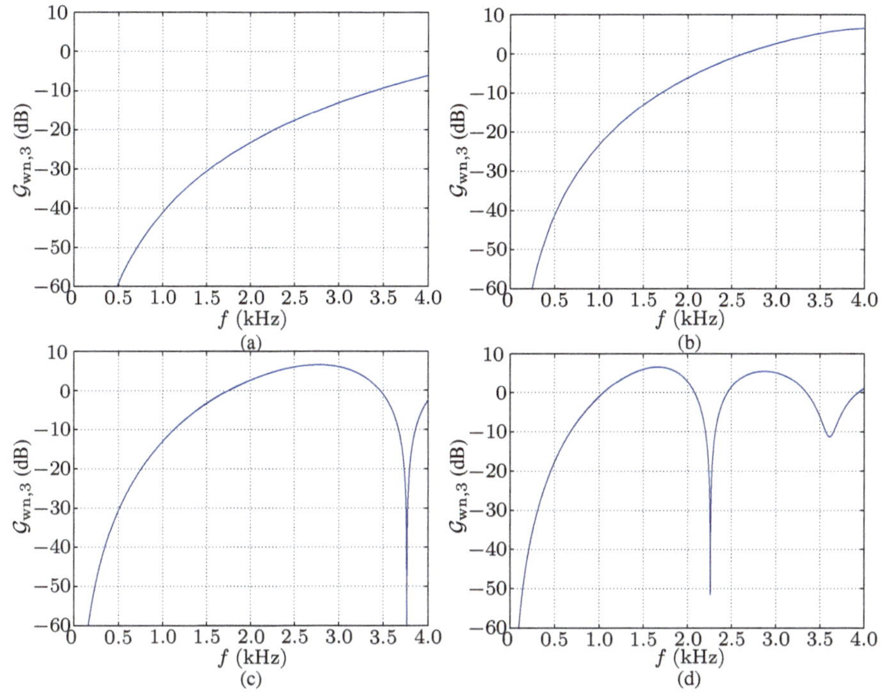


**Fig. 5.6** The white noise gain of the third-order CDMA (with seven microphones), as a function of frequency, for different values of $\delta$: (a) $\delta = 1$ cm, (b) $\delta = 2$ cm, (c) $\delta = 3$ cm, and (d) $\delta = 5$ cm.

## 5.3 Generalization

It is obvious from all what we have learned so far that a CDMA of order $\lfloor \frac{M}{2} \rfloor = N$ can be designed with $M$ sensors. In this context, it is clear that the system of linear equations to solve is

$$\mathbf{A}_M(\omega, \theta)\,\mathbf{h}(\omega) = \mathbf{b}_M, \tag{5.19}$$

where

$$\mathbf{A}_M(\omega, \theta) = \begin{bmatrix} \mathbf{d}^H(\omega, 0) \\ \mathbf{d}^H(\omega, \theta_{N,1}) \\ \vdots \\ \mathbf{d}^H(\omega, \theta_{N,N}) \\ \mathbf{c}_{M,1}^T \\ \vdots \\ \mathbf{c}_{M,M-N-1}^T \end{bmatrix} \tag{5.20}$$

is a square matrix of size $M \times M$, $\mathbf{c}_{M,i}$, $i = 1, 2, \ldots, M - N - 1$ are the constraints vectors of length $M$,

$$\mathbf{b}_M = \begin{bmatrix} 1 \ \beta_{N,1} \ \cdots \ \beta_{N,N} \ 0 \cdots 0 \end{bmatrix}^T, \tag{5.21}$$

$\theta_{N,n}$, $n = 1, 2, \ldots, N$ are given by design, $0 < \theta_{N,n} \leq \pi$, $\theta_{N,1} \neq \theta_{N,2} \neq \cdots \neq \theta_{N,N}$, and $0 \leq \beta_{N,n} \leq 1$. Since $\mathbf{A}_M (\omega, \theta)$ is a full-rank square matrix, we deduce that any $N$th-order CDMA can be designed by the filter:

$$\mathbf{h}(\omega) = \mathbf{A}_M^{-1} (\omega, \theta) \, \mathbf{b}_M. \tag{5.22}$$

We conclude this chapter by stating that we can design any third-order CDMA with six or seven microphones. However, the directivity factor is not that much improved as compared to second-order CDMAs, while white noise amplification can be much worse.

## References

1. J. Benesty and J. Chen, *Study and Design of Differential Microphone Arrays*. Berlin, Germany: Springer-Verlag, 2012.

# Chapter 6
# Superdirective Beamforming with Circular Arrays

In this chapter, we derive all kind of superdirective beamformers for both linear and circular arrays. We also show how these beamformers are strongly related to DMAs as both approaches lead to large array gains, i.e., supergains.

## 6.1 Superdirective Beamforming with ULAs

We consider a ULA consisting of $M$ omnidirectional microphones and for which the interelement spacing is $\delta$. As explained in Chapter 2, the steering vector associated with this antenna geometry is $\mathbf{d}_{\mathrm{L}}(\omega, \theta)$. It is well known that for a small value of $\delta$ and in the presence of a spherically isotropic noise, we can design a superdirective beamformer, $\mathbf{h}(\omega)$, at the endfire whose array gain approaches $M^2$ [1].

Within this context, the array gain is defined as

$$\mathcal{G}_{\mathrm{L,dn}}[\mathbf{h}(\omega)] = \frac{\left|\mathbf{h}^H(\omega)\,\mathbf{d}_{\mathrm{L}}(\omega,0)\right|^2}{\mathbf{h}^H(\omega)\,\mathbf{\Gamma}_{\mathrm{dn}}(\omega)\,\mathbf{h}(\omega)}, \tag{6.1}$$

where

$$[\mathbf{\Gamma}_{\mathrm{dn}}(\omega)]_{ij} = \mathrm{sinc}\left(\frac{\omega\delta_{ij}}{c}\right) \tag{6.2}$$

and

$$\delta_{ij} = (i-j)\delta. \tag{6.3}$$

It is obvious that the filter that maximizes $\mathcal{G}_{\mathrm{L,dn}}[\mathbf{h}(\omega)]$ is given by

$$\mathbf{h}_{\mathrm{L,max}}(\omega) = \varsigma(f)\mathbf{\Gamma}_{\mathrm{dn}}^{-1}(\omega)\,\mathbf{d}_{\mathrm{L}}(\omega,0), \tag{6.4}$$

 
J. Benesty et al., *Design of Circular Differential Microphone Arrays*,
Springer Topics in Signal Processing 12, DOI 10.1007/978-3-319-14842-7_6

where $\varsigma(f) \neq 0$ is an arbitrary complex number. Using the conventional distortionless constraint, i.e., $\mathbf{h}^H(\omega)\mathbf{d}_\mathrm{L}(\omega,0) = 1$, we deduce that the maximum SNR filter is

$$\mathbf{h}_{\mathrm{L,max}}(\omega) = \frac{\mathbf{\Gamma}_{\mathrm{dn}}^{-1}(\omega)\mathbf{d}_\mathrm{L}(\omega,0)}{\mathbf{d}_\mathrm{L}^H(\omega,0)\mathbf{\Gamma}_{\mathrm{dn}}^{-1}(\omega)\mathbf{d}_\mathrm{L}(\omega,0)} \tag{6.5}$$

and the corresponding gain is

$$\mathcal{G}_{\mathrm{L,dn}}[\mathbf{h}_{\mathrm{L,max}}(\omega)] = \mathbf{d}_\mathrm{L}^H(\omega,0)\mathbf{\Gamma}_{\mathrm{dn}}^{-1}(\omega)\mathbf{d}_\mathrm{L}(\omega,0). \tag{6.6}$$

The filter $\mathbf{h}_{\mathrm{L,max}}(\omega)$ is called superdirective beamformer. It can be shown that [2]

$$\lim_{\delta\to 0}\mathcal{G}_{\mathrm{L,dn}}[\mathbf{h}_{\mathrm{L,max}}(\omega)] = M^2. \tag{6.7}$$

Actually, for $\delta$ small, the filter $\mathbf{h}_{\mathrm{L,max}}(\omega)$ corresponds to the $(M-1)$th-order hypercardioid pattern since the latter is, by its own definition, obtained from the maximization of the directivity factor [3]. This shows how superdirective beamforming is strongly related to DMAs.

While with this approach we can achieve interesting SNR gains (as far as diffuse noise is concerned), the huge drawback that limits its use in practice is white noise amplification. Intuitively, it is easy to see that when $M$ is large and $\delta$ is small, $\mathbf{\Gamma}_{\mathrm{dn}}(\omega)$ becomes a very ill-conditioned matrix. As a result, the condition number of $\mathbf{\Gamma}_{\mathrm{dn}}(\omega)$ is very large and any small error in the system is largely amplified in $\mathbf{h}_{\mathrm{L,max}}(\omega)$.

It is not hard to observe that the white noise gain is

$$\begin{aligned}\mathcal{G}_{\mathrm{L,wn}}[\mathbf{h}_{\mathrm{L,max}}(\omega)] &= \frac{1}{\mathbf{h}_{\mathrm{L,max}}^H(\omega)\mathbf{h}_{\mathrm{L,max}}(\omega)} \\ &= \frac{\mathcal{G}_{\mathrm{L,dn}}^2[\mathbf{h}_{\mathrm{L,max}}(\omega)]}{\mathbf{d}_\mathrm{L}^H(\omega,0)\mathbf{\Gamma}_{\mathrm{dn}}^{-2}(\omega)\mathbf{d}_\mathrm{L}(\omega,0)} \leq M.\end{aligned} \tag{6.8}$$

In fact, in low frequencies, $\mathcal{G}_{\mathrm{L,wn}}[\mathbf{h}_{\mathrm{L,max}}(\omega)]$ can be much smaller than 1 and this implies that spatial white noise is amplified. In order to deal with this important problem, the authors in [1], [4] proposed to maximize the directivity factor, $\mathcal{G}_{\mathrm{L,dn}}[\mathbf{h}(\omega)]$, subject to a constraint on the white noise gain, $\mathcal{G}_{\mathrm{L,wn}}[\mathbf{h}(\omega)]$. This is equivalent to minimizing $1/\mathcal{G}_{\mathrm{L,dn}}[\mathbf{h}(\omega)]$ with a constraint on $1/\mathcal{G}_{\mathrm{L,wn}}[\mathbf{h}(\omega)]$, i.e., minimizing

$$\frac{1}{\mathcal{G}_{\mathrm{L,dn}}[\mathbf{h}(\omega)]} + \epsilon\frac{1}{\mathcal{G}_{\mathrm{L,wn}}[\mathbf{h}(\omega)]} = \frac{\mathbf{h}^H(\omega)[\mathbf{\Gamma}_{\mathrm{dn}}(\omega) + \epsilon\mathbf{I}_M]\mathbf{h}(\omega)}{|\mathbf{h}^H(\omega)\mathbf{d}_\mathrm{L}(\omega,0)|^2}, \tag{6.9}$$

where $\epsilon$ is a Lagrange multiplier. Using the distortionless constraint, we easily find that the optimal solution is

$$\mathbf{h}_{\mathrm{L},\epsilon}(\omega) = \frac{\left[\mathbf{\Gamma}_{\mathrm{dn}}(\omega) + \epsilon \mathbf{I}_M\right]^{-1} \mathbf{d}_{\mathrm{L}}(\omega, 0)}{\mathbf{d}_{\mathrm{L}}^H(\omega, 0) \left[\mathbf{\Gamma}_{\mathrm{dn}}(\omega) + \epsilon \mathbf{I}_M\right]^{-1} \mathbf{d}_{\mathrm{L}}(\omega, 0)}. \tag{6.10}$$

This filter is called robust superdirective beamformer. It is clear that (6.10) is a regularized version of (6.5), where $\epsilon$ is the regularization parameter. This parameter tries to find a good compromise between a supergain and white noise amplification. A small $\epsilon$ leads to a large directivity factor and a low white noise gain, while a large $\epsilon$ gives a low directivity factor and a large white noise gain. Obviously, we have

$$\mathbf{h}_{\mathrm{L},0}(\omega) = \mathbf{h}_{\mathrm{L},\max}(\omega) \tag{6.11}$$

and

$$\mathbf{h}_{\mathrm{L},\infty}(\omega) = \frac{\mathbf{d}_{\mathrm{L}}(\omega, 0)}{M}, \tag{6.12}$$

which corresponds to the maximum possible white noise gain, i.e.,

$$\mathcal{G}_{\mathrm{L,wn}}\left[\mathbf{h}_{\mathrm{L},\infty}(\omega)\right] = M. \tag{6.13}$$

However,

$$\mathcal{G}_{\mathrm{L,dn}}\left[\mathbf{h}_{\mathrm{L},\infty}(\omega)\right] = \frac{M^2}{\mathbf{d}_{\mathrm{L}}^H(\omega, 0)\, \mathbf{\Gamma}_{\mathrm{dn}}(\omega)\, \mathbf{d}_{\mathrm{L}}(\omega, 0)} < M^2. \tag{6.14}$$

In (6.10), it may be better to make $\epsilon$ frequency dependent since in low frequencies, the regularization parameter should be large, while in high frequencies, it should be small.

Before leaving this section, it is important to notice that the beampattern corresponding to $\mathbf{h}_{\mathrm{L},\epsilon}(\omega)$ is symmetric about the axis $0; \pi$, i.e.,

$$\begin{aligned} \mathcal{B}_{\mathrm{L}}\left[\mathbf{h}_{\mathrm{L},\epsilon}(\omega), \theta\right] &= \mathbf{h}_{\mathrm{L},\epsilon}^H(\omega)\, \mathbf{d}_{\mathrm{L}}(\omega, \theta) \\ &= \mathcal{B}_{\mathrm{L}}\left[\mathbf{h}_{\mathrm{L},\epsilon}(\omega), -\theta\right]. \end{aligned} \tag{6.15}$$

## 6.2 Superdirective Beamforming with UCAs

In this section, we study three different forms of superdirective beamformers with a circular array. We assume that the UCA is of radius $r$ and consists of $M$ omnidirectional microphones. We also often assume that $\theta_{\mathrm{s}} = 0$. We know from Chapter 2 that, by permuting the coefficients of the designed filter, we can easily steer the array at the angles $\theta_{\mathrm{s}} = \psi_m,\ m = 1, 2, \ldots, M$, where $\psi_m$ is the angular position of the $m$th microphone. We also know from Chapter 2 that the array gain in diffuse noise is

$$\mathcal{G}_{\mathrm{dn}}\left[\mathbf{h}(\omega)\right]=\frac{\left|\mathbf{h}^{H}(\omega)\,\mathbf{d}(\omega,0)\right|^{2}}{\mathbf{h}^{H}(\omega)\,\boldsymbol{\Gamma}_{\mathrm{dn}}(\omega)\,\mathbf{h}(\omega)},\tag{6.16}$$

where $\mathbf{d}(\omega,0)$ and $\boldsymbol{\Gamma}_{\mathrm{dn}}(\omega)$ are already defined in Chapter 2.

### *6.2.1 Without the Symmetry Constraint*

In this subsection, we do not impose any constraint on the filter $\mathbf{h}(\omega)$. As a result, $\mathcal{B}\left[\mathbf{h}(\omega),\theta\right]\neq\mathcal{B}\left[\mathbf{h}(\omega),-\theta\right]$, i.e., the beampattern corresponding to $\mathbf{h}(\omega)$ has no symmetry at $\theta_{\mathrm{s}}=0$.

By following the same steps as in Section 6.1, we find that the superdirective beamformer is [5]

$$\mathbf{h}_{\mathrm{max}}(\omega)=\frac{\boldsymbol{\Gamma}_{\mathrm{dn}}^{-1}(\omega)\,\mathbf{d}(\omega,0)}{\mathbf{d}^{H}(\omega,0)\,\boldsymbol{\Gamma}_{\mathrm{dn}}^{-1}(\omega)\,\mathbf{d}(\omega,0)}\tag{6.17}$$

and the corresponding gain is

$$\mathcal{G}_{\mathrm{dn}}\left[\mathbf{h}_{\mathrm{max}}(\omega)\right]=\mathbf{d}^{H}(\omega,0)\,\boldsymbol{\Gamma}_{\mathrm{dn}}^{-1}(\omega)\,\mathbf{d}(\omega,0).\tag{6.18}$$

*Property 6.1.* The coefficients of $\mathbf{h}_{\mathrm{max}}(\omega)$ have the following symmetry:

$$H_{m+1,\mathrm{max}}(\omega)=H_{M-m+1,\mathrm{max}}(\omega),\ m=1,2,\ldots,M-1.\tag{6.19}$$

This is quite remarkable and it is a consequence of the double coincidence that $\boldsymbol{\Gamma}_{\mathrm{dn}}(\omega)$ is circulant (and symmetric) and the elements of $\mathbf{d}(\omega,0)$ have a symmetry.

*Proof.* Since $\boldsymbol{\Gamma}_{\mathrm{dn}}(\omega)$ is circulant (Property 2.4) and symmetric, so is $\boldsymbol{\Gamma}_{\mathrm{dn}}^{-1}(\omega)$ [6]. Then, it is not hard to show that the multiplication of a symmetric circulant matrix with a vector whose coefficients have the symmetry given in (2.9) gives necessarily a vector whose coefficients have the same symmetry, i.e., (6.19).

In the general case:

$$\mathbf{h}_{\mathrm{max}}(\omega)=\frac{\boldsymbol{\Gamma}_{\mathbf{v}}^{-1}(\omega)\,\mathbf{d}(\omega,0)}{\mathbf{d}^{H}(\omega,0)\,\boldsymbol{\Gamma}_{\mathbf{v}}^{-1}(\omega)\,\mathbf{d}(\omega,0)},\tag{6.20}$$

where $\boldsymbol{\Gamma}_{\mathbf{v}}^{-1}(\omega)$ is not circulant, $\mathbf{h}_{\mathrm{max}}(\omega)$ has no particular symmetry.

*Property 6.2.* By just permuting the coefficients of $\mathbf{h}_{\mathrm{max}}(\omega)$, the array can be steered in the $M$ different angles $\psi_m$, $m=1,2,\ldots,M$, with the same symmetric beampattern.

*Proof.* Since $\mathbf{\Gamma}_{\mathrm{dn}}^{-1}(\omega)$ is a circulant matrix, it can be written as

$$\mathbf{\Gamma}_{\mathrm{dn}}^{-1}(\omega) = \begin{bmatrix} c_1(\omega) & c_M(\omega) & \cdots & c_3(\omega) & c_2(\omega) \\ c_2(\omega) & c_1(\omega) & \cdots & c_4(\omega) & c_3(\omega) \\ \vdots & \vdots & \ddots & \vdots & \vdots \\ c_{M-1}(\omega) & c_{M-2}(\omega) & \cdots & c_1(\omega) & c_M(\omega) \\ c_M(\omega) & c_{M-1}(\omega) & \cdots & c_2(\omega) & c_1(\omega) \end{bmatrix}, \tag{6.21}$$

where $c_{M-m+1}(\omega) = c_{m+1}(\omega)$, $m = 1, 2, \ldots, M-1$ since $\mathbf{\Gamma}_{\mathrm{dn}}^{-1}(\omega)$ is symmetric. Using the permutation matrices, we can express (6.21) as

$$\mathbf{\Gamma}_{\mathrm{dn}}^{-1}(\omega) = \sum_{m=1}^{M} c_m(\omega)\mathbf{P}_m = \sum_{m=1}^{M} c_m(\omega)\mathbf{P}_2^{m-1}. \tag{6.22}$$

Furthermore, we easily derive

$$\begin{aligned} \mathbf{P}_m \mathbf{\Gamma}_{\mathrm{dn}}^{-1}(\omega)\mathbf{P}_m^T &= \mathbf{P}_m \left[ \sum_{i=1}^{M} c_i(\omega)\mathbf{P}_2^{i-1} \right] \mathbf{P}_m^T \\ &= \mathbf{P}_2^{m-1} \left[ \sum_{i=1}^{M} c_i(\omega)\mathbf{P}_2^{i-1} \right] \left(\mathbf{P}_2^{m-1}\right)^T \\ &= \sum_{i=1}^{M} c_i(\omega) \left[ \mathbf{P}_2^{m-1}\mathbf{P}_2^{i-1} \left(\mathbf{P}_2^{m-1}\right)^T \right] \\ &= \sum_{i=1}^{M} c_i(\omega)\mathbf{P}_2^{i-1} \\ &= \mathbf{\Gamma}_{\mathrm{dn}}^{-1}(\omega). \end{aligned} \tag{6.23}$$

Therefore, we have

$$\begin{aligned} \mathbf{P}_m \mathbf{h}_{\max}(\omega) &= \frac{\mathbf{P}_m \mathbf{\Gamma}_{\mathrm{dn}}^{-1}(\omega)\mathbf{d}(\omega, 0)}{\mathbf{d}^H(\omega, 0)\mathbf{\Gamma}_{\mathrm{dn}}^{-1}(\omega)\mathbf{d}(\omega, 0)} \\ &= \frac{\mathbf{P}_m \mathbf{\Gamma}_{\mathrm{dn}}^{-1}(\omega)\mathbf{P}_m^T\mathbf{d}(\omega, \psi_m)}{\mathbf{d}^H(\omega, \psi_m)\mathbf{P}_m \mathbf{\Gamma}_{\mathrm{dn}}^{-1}(\omega)\mathbf{P}_m^T\mathbf{d}(\omega, \psi_m)} \\ &= \frac{\mathbf{\Gamma}_{\mathrm{dn}}^{-1}(\omega)\mathbf{d}(\omega, \psi_m)}{\mathbf{d}^H(\omega, \psi_m)\mathbf{\Gamma}_{\mathrm{dn}}^{-1}(\omega)\mathbf{d}(\omega, \psi_m)} \\ &= \mathbf{h}_{\max}(\omega, \psi_m). \end{aligned} \tag{6.24}$$

The beampattern of $\mathbf{h}_{\max}(\omega, \theta_{\mathrm{s}})$ for $\theta_{\mathrm{s}} \neq \psi_m$ may be symmetric but it will be very different from the beampattern of $\mathbf{h}_{\max}(\omega, \psi_m)$.

Since the order of a superdirective UCA beamformer is $\lfloor \frac{M}{2} \rfloor$ and by analogy to a superdirective ULA beamformer, we should have[1]

$$\lim_{\delta \to 0} \mathcal{G}_{\text{dn}} \left[ \mathbf{h}_{\max} (\omega) \right] \approx \left( \lfloor \frac{M}{2} \rfloor + 1 \right)^2 . \tag{6.25}$$

For $r$ (or $\delta$) small, the superdirective beamformer given in (6.17) corresponds to the hypercardioid pattern of order $\lfloor \frac{M}{2} \rfloor$.

It is straightforward to derive the robust superdirective beamformer [5]:

$$\mathbf{h}_{\epsilon} (\omega) = \frac{\left[ \mathbf{\Gamma}_{\text{dn}} (\omega) + \epsilon \mathbf{I}_M \right]^{-1} \mathbf{d} (\omega, 0)}{\mathbf{d}^H (\omega, 0) \left[ \mathbf{\Gamma}_{\text{dn}} (\omega) + \epsilon \mathbf{I}_M \right]^{-1} \mathbf{d} (\omega, 0)}. \tag{6.26}$$

We have

$$\mathbf{h}_0 (\omega) = \mathbf{h}_{\max} (\omega) \tag{6.27}$$

and

$$\mathbf{h}_{\infty} (\omega) = \frac{\mathbf{d} (\omega, 0)}{M}, \tag{6.28}$$

which corresponds to the maximum possible white noise gain, i.e.,

$$\mathcal{G}_{\text{wn}} \left[ \mathbf{h}_{\infty} (\omega) \right] = M. \tag{6.29}$$

## 6.2.2 With the Symmetry Constraint

In the first part of this subsection, we show how to obtain $\mathbf{h}_{\max} (\omega)$ from a different perspective. This approach can be seen as a generalization for the derivation of a symmetric beampattern whose shape is not affected even if the pseudo-coherence matrix of the noise matrix is no longer circulant.

To have a symmetric pattern at $\theta_{\text{s}} = 0$, we must use the relations on the coefficients of the filter (see Chapter 2):

$$H_{m+1} (\omega) = H_{M-m+1} (\omega) , \ m = 1, 2, \ldots, M - 1. \tag{6.30}$$

We deduce that we have $M' = M - \lfloor \frac{M}{2} \rfloor - 1$ constraints of the form:

$$\mathbf{c}_{M,m'}^T \mathbf{h} (\omega) = 0, \ m' = 1, 2, \ldots, M', \tag{6.31}$$

where $\mathbf{c}_{M,m'}$ is a vector of length $M$ whose $(m' + 1)$th element is 1, $(M - m' + 1)$th element is $-1$, and all its other components are 0. For example,

[1] This is true only if $M$ is not too large.

$$\mathbf{c}_{M,1} = \begin{bmatrix} 0 \ 1 \ 0 \cdots 0 \ -1 \end{bmatrix}^T, \tag{6.32}$$

$$\mathbf{c}_{M,2} = \begin{bmatrix} 0 \ 0 \ 1 \ 0 \cdots 0 \ -1 \ 0 \end{bmatrix}^T. \tag{6.33}$$

Combining together the constraints given in (6.31) and the distortionless constraint, we get

$$\mathbf{C}(\omega)\mathbf{h}(\omega) = \mathbf{i}_1, \tag{6.34}$$

where

$$\mathbf{C}(\omega) = \begin{bmatrix} \mathbf{d}^H(\omega, 0) \\ \mathbf{c}_{M,1}^T \\ \mathbf{c}_{M,2}^T \\ \vdots \\ \mathbf{c}_{M,M'}^T \end{bmatrix} \tag{6.35}$$

is a matrix of size $(M'+1) \times M$ and

$$\mathbf{i}_1 = \begin{bmatrix} 1 \ 0 \cdots 0 \end{bmatrix}^T \tag{6.36}$$

is a vector of length $M'+1$. Now, our optimization problem is the following:

$$\min_{\mathbf{h}(\omega)} \mathbf{h}^H(\omega)\mathbf{\Gamma}_{\mathrm{dn}}(\omega)\mathbf{h}(\omega) \quad \text{subject to} \quad \mathbf{C}(\omega)\mathbf{h}(\omega) = \mathbf{i}_1, \tag{6.37}$$

for which the solution is the superdirective beamformer:

$$\mathbf{h}_{\mathrm{max}}(\omega) = \mathbf{\Gamma}_{\mathrm{dn}}^{-1}(\omega)\mathbf{C}^H(\omega)\left[\mathbf{C}(\omega)\mathbf{\Gamma}_{\mathrm{dn}}^{-1}(\omega)\mathbf{C}^H(\omega)\right]^{-1}\mathbf{i}_1. \tag{6.38}$$

The previous maximum SNR filter is valid only for $\theta_{\mathrm{s}} = 0$. For $\theta_{\mathrm{s}} = \psi_m$, we just need to permute the coefficients of $\mathbf{h}_{\mathrm{max}}(\omega)$. We also deduce that

$$\mathcal{G}_{\mathrm{dn}}\left[\mathbf{h}_{\mathrm{max}}(\omega)\right] = \frac{1}{\mathbf{i}_1^T\left[\mathbf{C}(\omega)\mathbf{\Gamma}_{\mathrm{dn}}^{-1}(\omega)\mathbf{C}^H(\omega)\right]^{-1}\mathbf{i}_1}. \tag{6.39}$$

For $r$ (or $\delta$) small, the superdirective beamformer given in (6.38) corresponds to the hypercardioid pattern of order $\lfloor \frac{M}{2} \rfloor$.

We can derive the robust superdirective beamformer:

$$\mathbf{h}_{\epsilon}(\omega) = \left[\mathbf{\Gamma}_{\mathrm{dn}}(\omega) + \epsilon\mathbf{I}_M\right]^{-1}\mathbf{C}^H(\omega)\left\{\mathbf{C}(\omega)\left[\mathbf{\Gamma}_{\mathrm{dn}}(\omega) + \epsilon\mathbf{I}_M\right]^{-1}\mathbf{C}^H(\omega)\right\}^{-1}\mathbf{i}_1. \tag{6.40}$$

We have

$$\mathbf{h}_0(\omega) = \mathbf{h}_{\mathrm{max}}(\omega) \tag{6.41}$$

and

$$\mathbf{h}_{\infty}(\omega) = \mathbf{C}^H(\omega)\left[\mathbf{C}(\omega)\mathbf{C}^H(\omega)\right]^{-1}\mathbf{i}_1, \tag{6.42}$$

which is the minimum-norm solution of (6.34). The white noise gain for $\mathbf{h}_{\infty}(\omega)$ is

$$\mathcal{G}_{\rm wn}\left[\mathbf{h}_{\infty}(\omega)\right] = \frac{1}{\mathbf{i}_1^T\left[\mathbf{C}(\omega)\mathbf{C}^H(\omega)\right]^{-1}\mathbf{i}_1}. \tag{6.43}$$

### 6.2.3 With the Maximum of Nulls

An interesting superdirective beamformer that has a symmetry by its own design is the following. We wish to have a one at the angle $\psi_1 = 0$ and nulls at the angles $\psi_i$, $i = 2, 3, \ldots, M$. Therefore, the number of constraints is equal to the number of microphones and the linear system to solve is

$$\mathbf{C}_{\rm N}(\omega)\mathbf{h}(\omega) = \mathbf{i}_{M,1}, \tag{6.44}$$

where

$$\mathbf{C}_{\rm N}(\omega) = \begin{bmatrix} \mathbf{d}^H(\omega, 0) \\ \mathbf{d}^H(\omega, \psi_2) \\ \vdots \\ \mathbf{d}^H(\omega, \psi_M) \end{bmatrix} \tag{6.45}$$

is a constraint matrix of size $M \times M$ and $\mathbf{i}_{M,1}$ is the first column of $\mathbf{I}_M$. Because of the symmetry, it can be checked that $\mathbf{C}_{\rm N}(\omega)$ is a circulant matrix. We deduce from (6.44) that the superdirective filter with the maximum of nulls is

$$\mathbf{h}_{\rm N}(\omega) = \mathbf{C}_{\rm N}^{-1}(\omega)\mathbf{i}_{M,1}. \tag{6.46}$$

By permuting the coefficients of $\mathbf{h}_{\rm N}(\omega)$, we can get the same beamformer at the directions $\theta_{\rm s} = \psi_m$, $m = 1, 2, \ldots, M$.

## 6.3 Design Examples

### 6.3.1 Superdirective Beamforming with a ULA

Let us consider a ULA as illustrated in Fig. 1.1 with six microphones, i.e., $M = 6$. Suppose that the signal of interest propagates from 0°, i.e., $\theta_{\rm s} = 0°$.

Figure 6.1 plots the patterns of the superdirective beamformer given in (6.5) for several frequencies and two values of $\delta$ (1 and 2 cm). It is seen that the patterns are similar (or slightly different) at different frequencies and spacing. The white noise gain and directivity factor are plotted in Fig. 6.2 as a function of frequency. It is observed that this directivity factor is close to 16 dB for both $\delta = 1$ cm and $\delta = 2$ cm. This corroborates with the theoretical analysis that the superdirective beamformer with a ULA can achieve a maximum director factor of $M^2$ (or 15.6 dB for $M = 6$) when spacing is small. One can see that there is some degradation in the directivity factor in low frequencies. The white noise gain of the superdirective beamformer is smaller than $-40$ dB for $\delta = 1$ cm and smaller than $-10$ dB for $\delta = 2$ cm. The negative white noise gain means that this beamformer suffers from white noise amplification. The smaller is the white noise gain, the more is the white noise amplification. Obviously, the white noise amplification is more significant in low frequencies.

With the great advancement of sensing and micro-electro-mechanical-systems (MEMS) technology, many microphones can nowadays have self noise less than 25 dB. With this type of sensors, microphone arrays can tolerate some degree of white noise amplification (say 10 dB). But we see from the previous example that the superdirective beamformer with small spacing (1 cm or 2 cm) has a white noise gain smaller than $-10$ dB, and in low frequencies this gain is even below $-100$ dB. This renders this beamformer impractical in real applications. One way to improve the white noise gain is via the robust superdirective beamformer as given in (6.10). With the same ULA, the patterns of the robust superdirective beamformer with $\epsilon = 0.001$ for several frequencies and two values of $\delta$ (1 and 2 cm) are plotted in Fig. 6.3, and the white noise gain and directivity factor are plotted in Fig. 6.4 as a function of frequency. We can see that the robust superdirective beamformer has a white noise gain close to $-10$ dB and this gain is almost frequency independent. However, the patterns of the beamformer vary with frequency and so is the directivity factor. Therefore, the robust superdirective beamformer achieves its robustness with respect to white noise at the price of losing the frequency independency of the beampattern.

### 6.3.2 Superdirective Beamforming with a UCA – without the Symmetry Constraint

Let us consider a UCA with six microphones, i.e., $M = 6$. Figure 6.5 plots the patterns of the superdirective beamformer given in (6.17) for two frequencies, two values of $\delta$ (1 and 2 cm), and two values of $\theta_s$ (0° and 15°). When $\theta_s = 0°$, the patterns are symmetric with respect to the axis 0°–180° as seen from Fig. 6.5(a)–(d). Also, the patterns for $\theta_s = 0°$ are similar (or slightly different) at different frequencies and spacing. It has been shown from theory that the patterns for $\theta_s = \psi_m$ [in our case with six microphones,

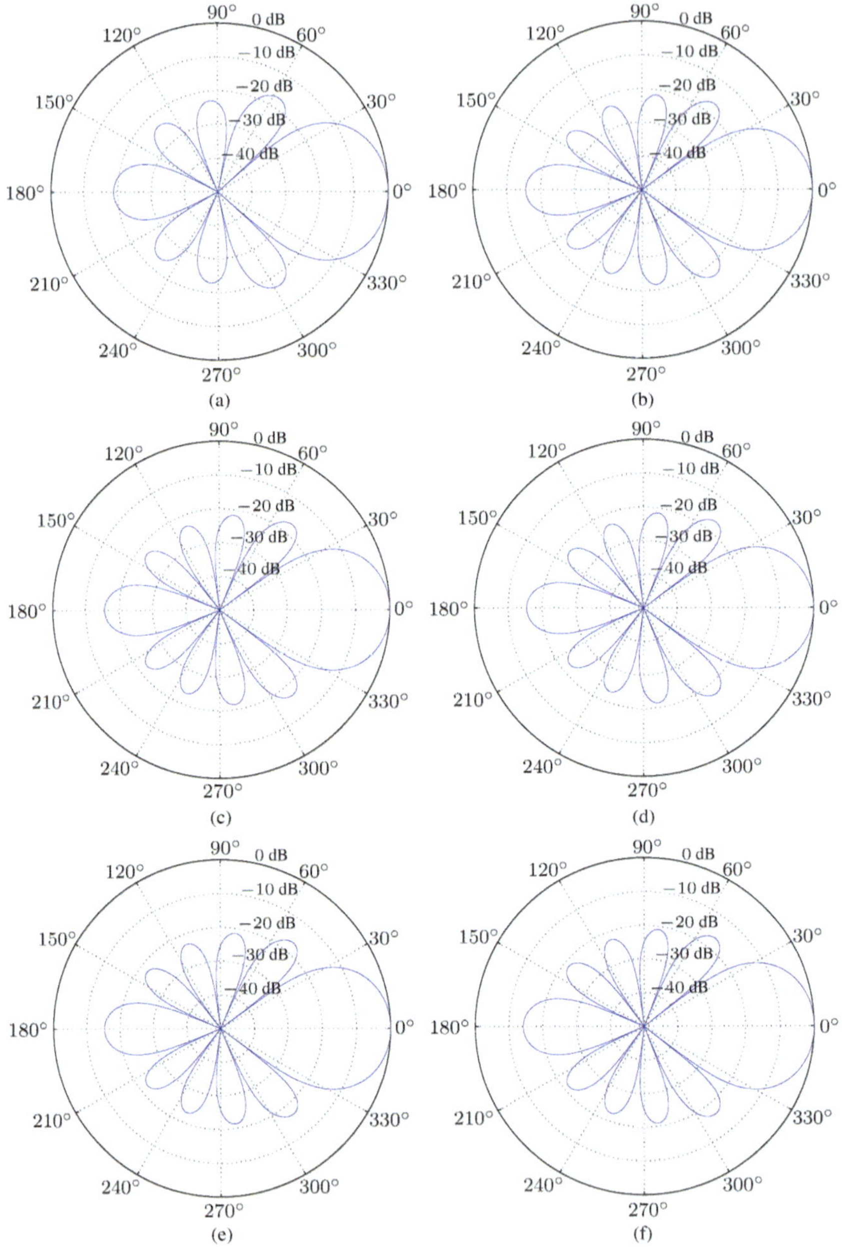


**Fig. 6.1** Patterns of the superdirective beamformer with a ULA ($M = 6$ and $\theta_s = 0°$) for several frequencies and two values of $\delta$: (a) $f = 0.5$ kHz, $\delta = 1$ cm, (b) $f = 0.5$ kHz, $\delta = 2$ cm, (c) $f = 1$ kHz, $\delta = 1$ cm, (d) $f = 1$ kHz, $\delta = 2$ cm, (e) $f = 3$ kHz, $\delta = 1$ cm, and (f) $f = 3$ kHz, $\delta = 2$ cm.

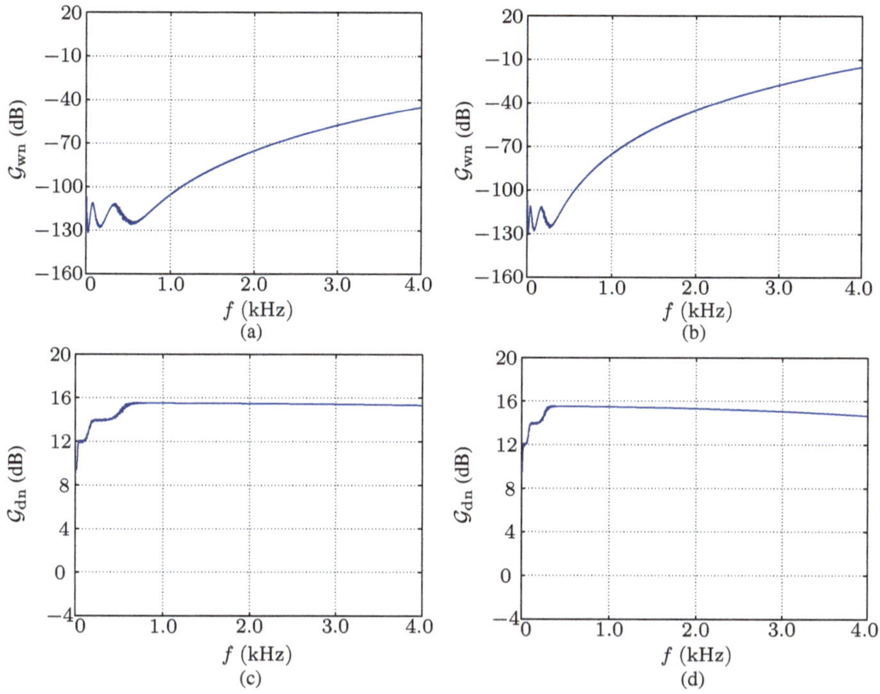


**Fig. 6.2** The white noise gain and directivity factor of the superdirective beamformer with a ULA ($M = 6$ and $\theta_s = 0°$) as a function of frequency for different values of $\delta$: (a) the white noise gain for $\delta = 1$ cm, (b) the white noise gain for $\delta = 2$ cm, (c) the directivity factor for $\delta = 1$ cm, and (d) the directivity factor for $\delta = 2$ cm.

$\psi_m = (m-1) \times 60°$, $m = 2, 3, \ldots, M$] are simply obtained from a rotation of patterns at $\theta_s = 0°$, so all the patterns for $\theta_s = \psi_m$ should also be symmetric (with respect to the line connecting $\psi_m$ and $180° + \psi_m$). However, if $\theta_s$ is not equal to one of the angles $\psi_m$, $m = 1, 2, \ldots, M$, the patterns are not guaranteed to be symmetric as seen from Figs. 6.5(e) and (f).

The white noise gain and directivity factor of the superdirective beamformer are plotted in Fig. 6.6 as a function of frequency with $\delta = 1$ cm. With this six-microphone UCA, the superdirective beamformer achieves a directivity factor close to 11 dB in most frequencies. Comparing Figs. 6.6 and 6.2, one can see that the directivity factor of the UCA is smaller than that of the ULA with the same number of sensors and same value of spacing. But the circular array has less white noise amplification as its white noise gain is much larger than that of the linear array. Besides the better white noise gain, another advantage of using a UCA is that it can produce the same beampattern in many different directions (at least for $\theta_s = \psi_m$, $m = 1, 2, \ldots, M$). For a ULA, however, the optimal performance of the superdirective beamformer occurs only at endfire directions [7].

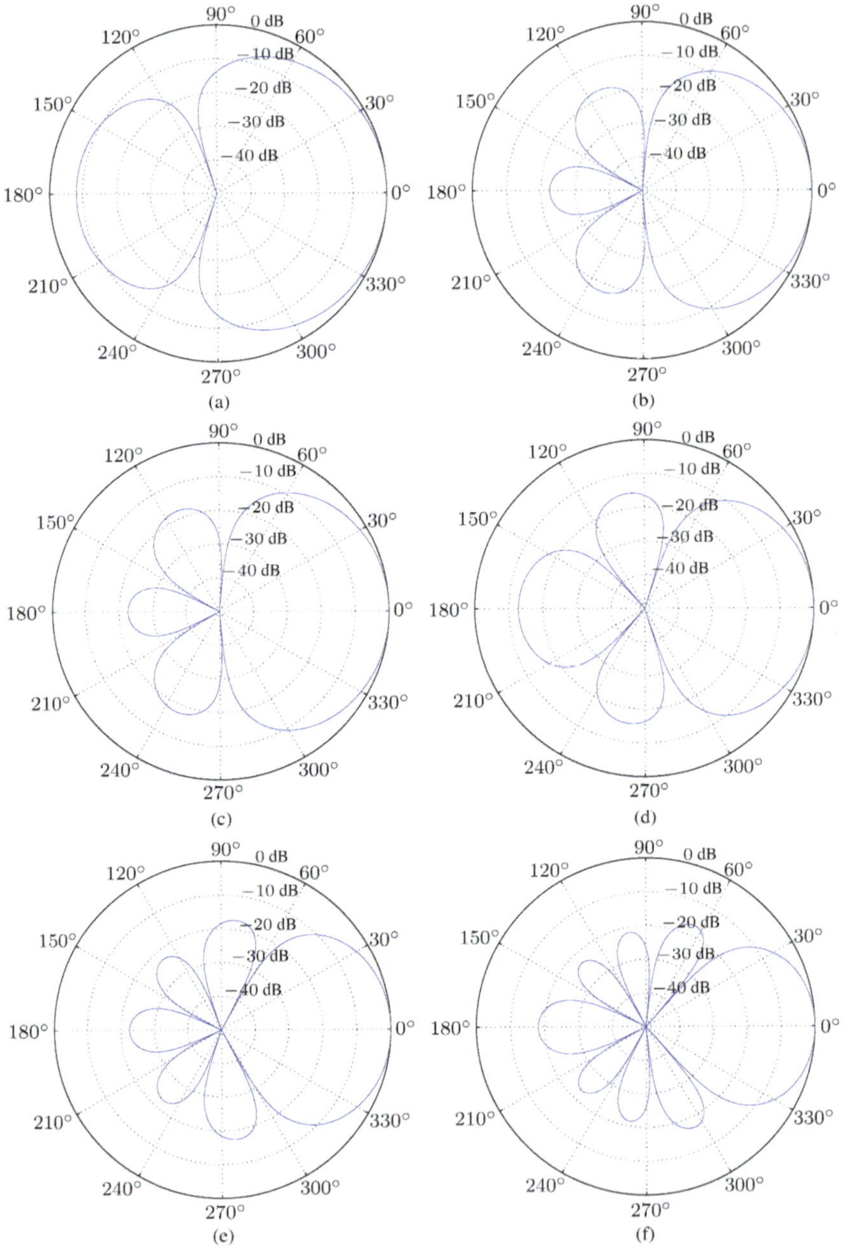


**Fig. 6.3** Patterns of the robust superdirective beamformer with a ULA ($M = 6$, $\theta_s = 0°$, and $\epsilon = 0.001$) for several frequencies and two values of $\delta$: (a) $f = 0.5$ kHz, $\delta = 1$ cm, (b) $f = 0.5$ kHz, $\delta = 2$ cm, (c) $f = 1$ kHz, $\delta = 1$ cm, (d) $f = 1$ kHz, $\delta = 2$ cm, (e) $f = 3$ kHz, $\delta = 1$ cm, and (f) $f = 3$ kHz, $\delta = 2$ cm.

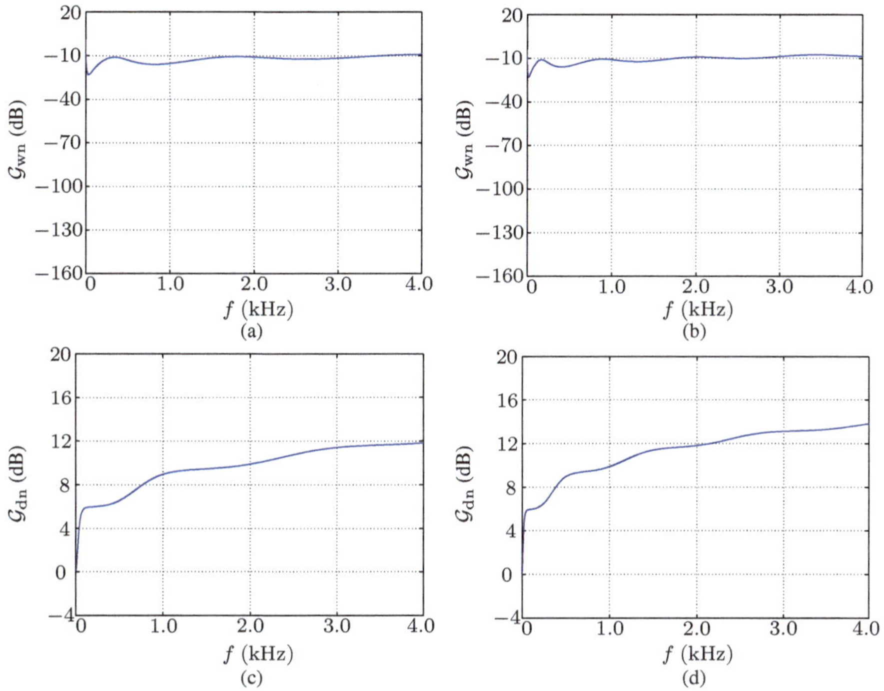


**Fig. 6.4** The white noise gain and directivity factor of the robust superdirective beamformer with a ULA ($M = 6$, $\theta_{\rm s} = 0°$, and $\epsilon = 0.001$) as a function of frequency for different values of $\delta$: (a) the white noise gain for $\delta = 1$ cm, (b) the white noise gain for $\delta = 2$ cm, (c) the directivity factor for $\delta = 1$ cm, and (d) the directivity factor for $\delta = 2$ cm.

Although it has a better white noise gain than the ULA with the same number of sensors and same spacing, the UCA still suffers from tremendously white noise amplification. Again, one can improve the problem by using the robust version of the superdirective beamformer as given in (6.26). With the same UCA, the patterns of the robust superdirective beamformer are plotted in Fig. 6.7 with $\epsilon = 0.001$, where we show the patterns for two frequencies, two values of $\delta$ (1 and 2 cm), and two values of $\theta_{\rm s}$ (0° and 15°). Now, we observe that the patterns change significantly with frequency and the value of spacing. The white noise gain and directivity factor corresponding to this robust superdirective beamformer are plotted in Fig. 6.8 as a function of frequency. We can see that this robust superdirective beamformer has a white noise gain close to $-10$ dB and this gain does not change much with frequency. Similar to the linear array situation, the improvement in white noise gain is achieved by sacrificing the directivity factor, which is clearly seen from Fig. 6.8, particularly in low frequencies.

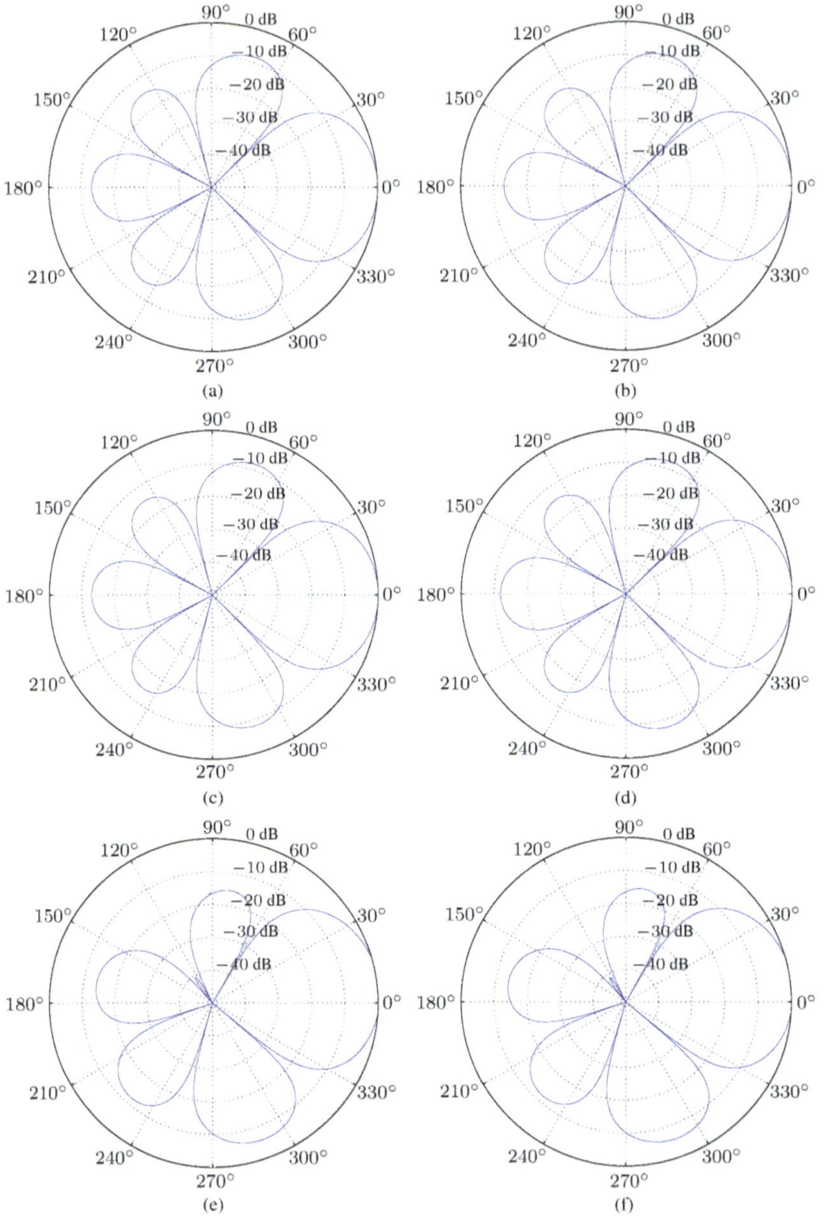


**Fig. 6.5** Patterns of the superdirective beamformer with a UCA ($M = 6$), without the symmetry constraint, for several frequencies, two values of $\delta$, and two values of $\theta_s$: (a) $f = 1$ kHz, $\delta = 1$ cm, $\theta_s = 0°$, (b) $f = 1$ kHz, $\delta = 2$ cm, $\theta_s = 0°$, (c) $f = 3$ kHz, $\delta = 1$ cm, $\theta_s = 0°$, (d) $f = 3$ kHz, $\delta = 2$ cm, $\theta_s = 0°$, (e) $f = 1$ kHz, $\delta = 1$ cm, $\theta_s = 15°$, and (f) $f = 1$ kHz, $\delta = 2$ cm, $\theta_s = 15°$.

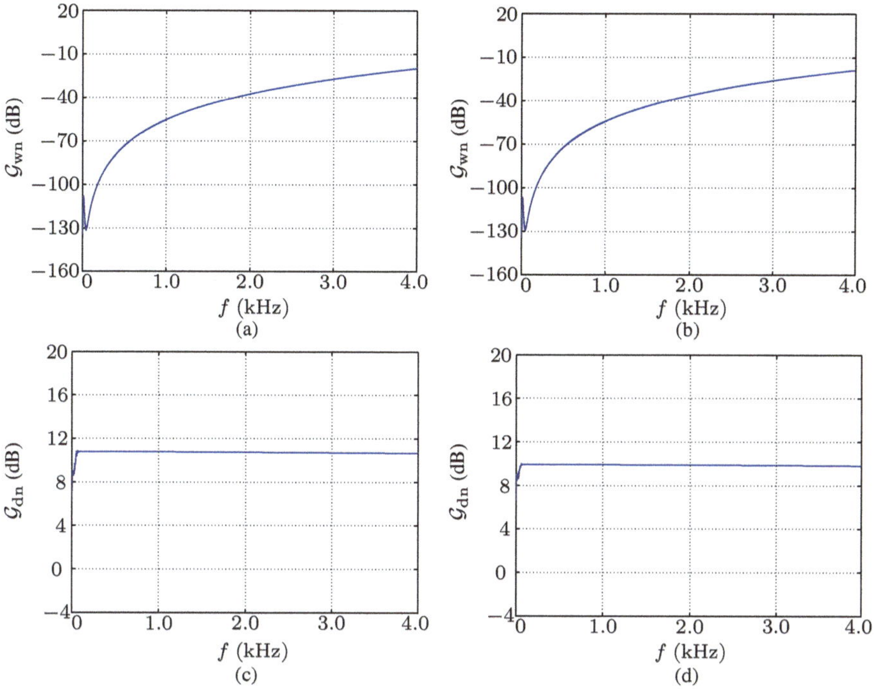


**Fig. 6.6** The white noise gain and directivity factor of the superdirective beamformer with a UCA ($M = 6$), without the symmetry constraint, as a function of frequency and two values of $\theta_s$: (a) the white noise gain for $\delta = 1$ cm, $\theta_s = 0°$, (b) the white noise gain for $\delta = 1$ cm, $\theta_s = 15°$, (c) the directivity factor for $\delta = 1$ cm, $\theta_s = 0°$, and (d) the directivity factor for $\delta = 1$ cm, $\theta_s = 15°$.

### *6.3.3 Superdirective Beamforming with a UCA – with the Symmetry Constraint*

Again, we consider a UCA with six microphones and study numerically the superdirective beamformer with the symmetry constraint. Figure 6.9 plots the patterns of the superdirective beamformer given in (6.38) for several frequencies and two values of $\delta$ (1 and 2 cm). It can be seen that all the patterns look almost the same. The white noise gain and directivity factor corresponding to this superdirective beamformer are plotted in Fig. 6.10 as a function of frequency. It is seen that this beamformer achieves a directivity factor close to 11 dB for both $\delta = 1$ cm and $\delta = 2$ cm. Similarly to the previous examples, the superdirective beamformer with the symmetry constraint also suffers from significant white noise amplification. But one can improve the white noise amplification problem by using the robust version of the superdirective beamformer as given in (6.40).

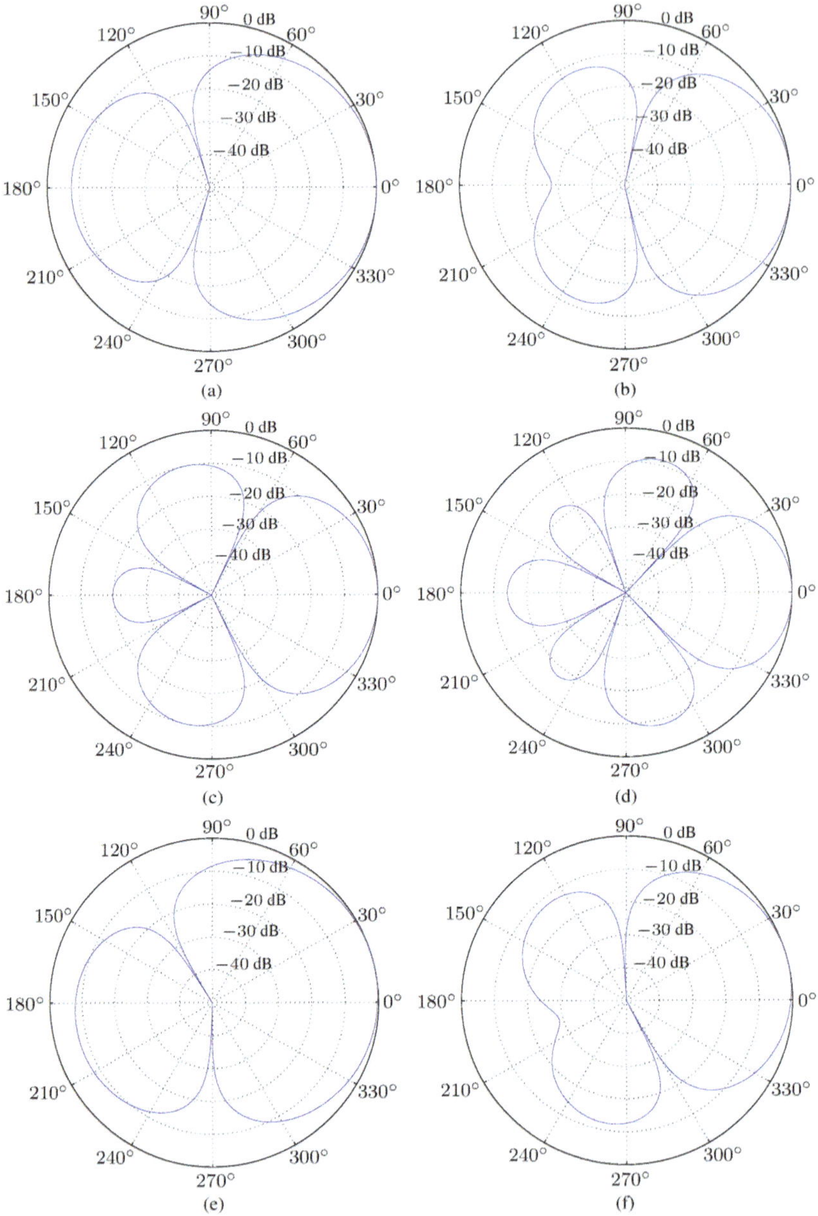


**Fig. 6.7** Patterns of the robust superdirective beamformer with a UCA ($M = 6$ and $\epsilon = 0.001$), without the symmetry constraint, for several frequencies, two values of $\delta$, and two values of $\theta_s$: (a) $f = 1$ kHz, $\delta = 1$ cm, $\theta_s = 0°$, (b) $f = 1$ kHz, $\delta = 2$ cm, $\theta_s = 0°$, (c) $f = 3$ kHz, $\delta = 1$ cm, $\theta_s = 0°$, (d) $f = 3$ kHz, $\delta = 2$ cm, $\theta_s = 0°$, (e) $f = 1$ kHz, $\delta = 1$ cm, $\theta_s = 15°$, and (f) $f = 1$ kHz, $\delta = 2$ cm, $\theta_s = 15°$.

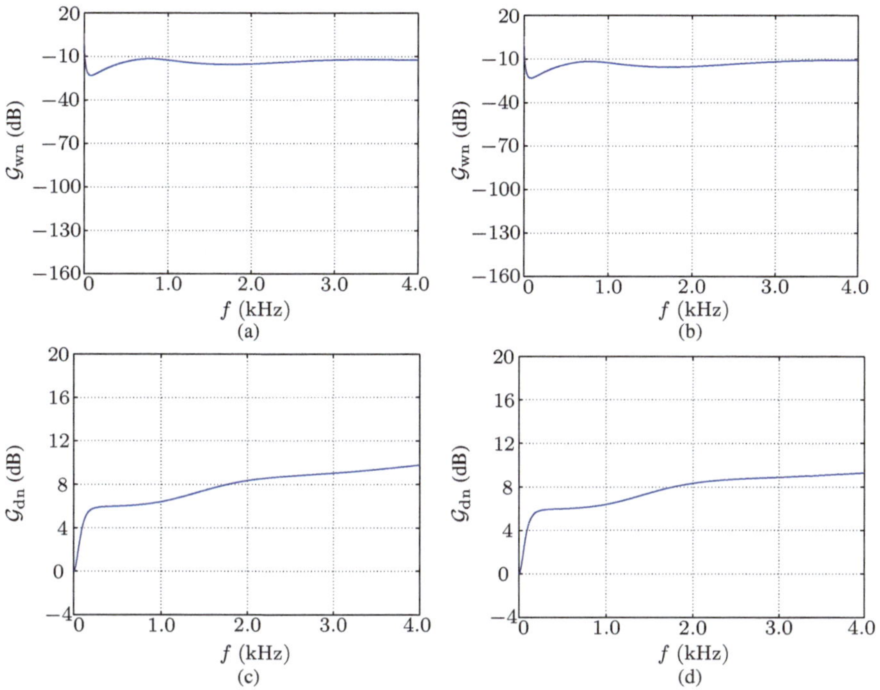


**Fig. 6.8** The white noise gain and directivity factor of the robust superdirective beamformer with a UCA ($M = 6$ and $\epsilon = 0.001$), without the symmetry constraint, as a function of frequency for two values of $\theta_{\mathrm{s}}$: (a) the white noise gain for $\delta = 1$ cm, $\theta_{\mathrm{s}} = 0°$, (b) the white noise gain for $\delta = 1$ cm, $\theta_{\mathrm{s}} = 15°$, (c) the directivity factor for $\delta = 1$ cm, $\theta_{\mathrm{s}} = 0°$, and (d) the directivity factor for $\delta = 1$ cm, $\theta_{\mathrm{s}} = 15°$.

### *6.3.4 Superdirective Beamforming with a UCA – with the Maximum of Nulls*

In this last example, we consider the superdirective beamformer with the maximum of nulls as discussed in Subsection 6.2.3. Again, we use a UCA with six microphones, i.e., $M = 6$. The patterns of the superdirective beamformer given in (6.46) are plotted in Fig 6.11 for several frequencies and two values of $\delta$ (1 and 2 cm). It is seen that all the patterns look the same and they all have 5 nulls at $\psi_i = i \times 60°$, $i = 1, 2, 3, 4, 5$. The white noise gain and directivity factor corresponding to this superdirective beamformer are plotted in Fig. 6.12 as a function of frequency. It is observed that the superdirective beamformer with the maximum of nulls achieves a directivity factor close to 10 dB for both $\delta = 1$ cm and $\delta = 2$ cm in most frequencies and the directivity factor does not change much with frequency. Similarly to the previous examples, the superdirective beamformer with the maximum of nulls suffers from some white noise amplification.

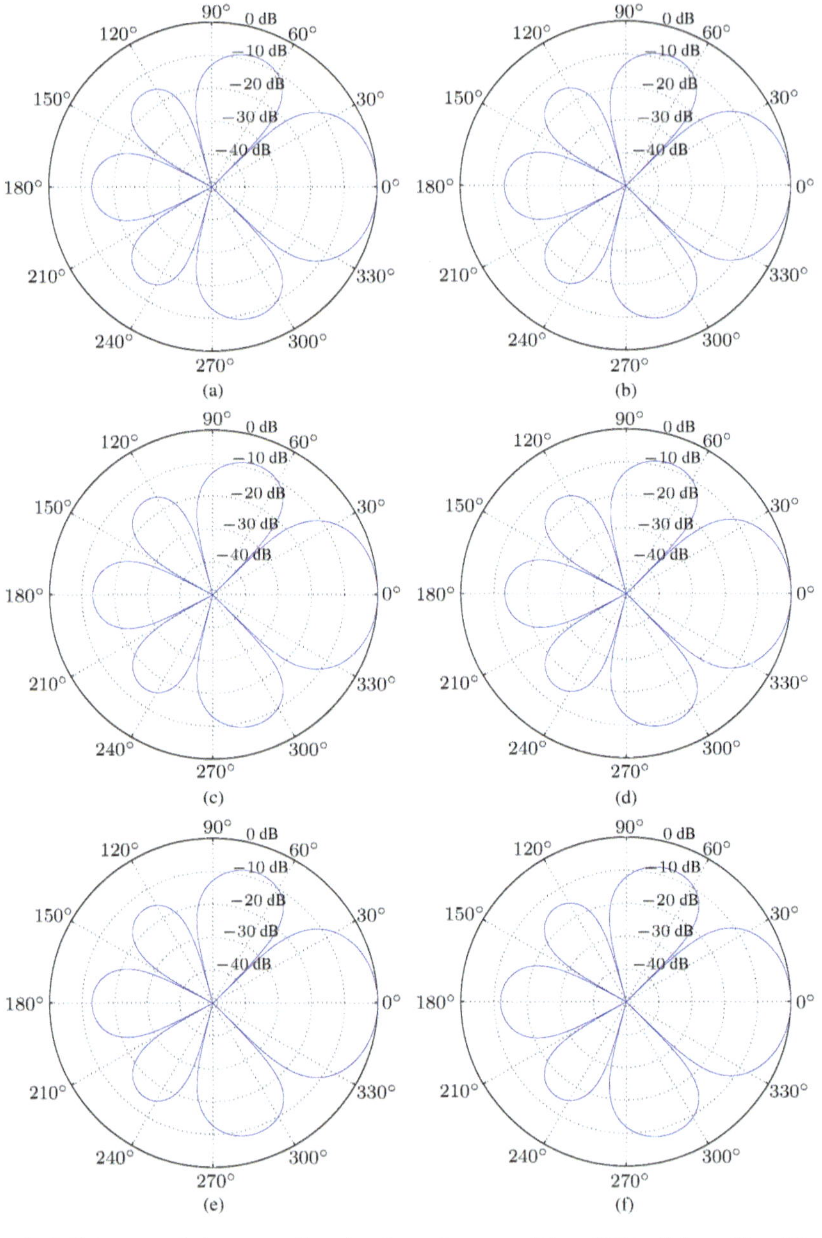


**Fig. 6.9** Patterns of the superdirective beamformer with a UCA ($M = 6$), with the symmetry constraint, for several frequencies, two values of $\delta$, and two values of $\theta_s = 0°$): (a) $f = 0.5$ kHz, $\delta = 1$ cm, (b) $f = 0.5$ kHz, $\delta = 2$ cm, (c) $f = 1$ kHz, $\delta = 1$ cm, (d) $f = 1$ kHz, $\delta = 2$ cm, (e) $f = 3$ kHz, $\delta = 1$ cm, and (f) $f = 3$ kHz, $\delta = 2$ cm.

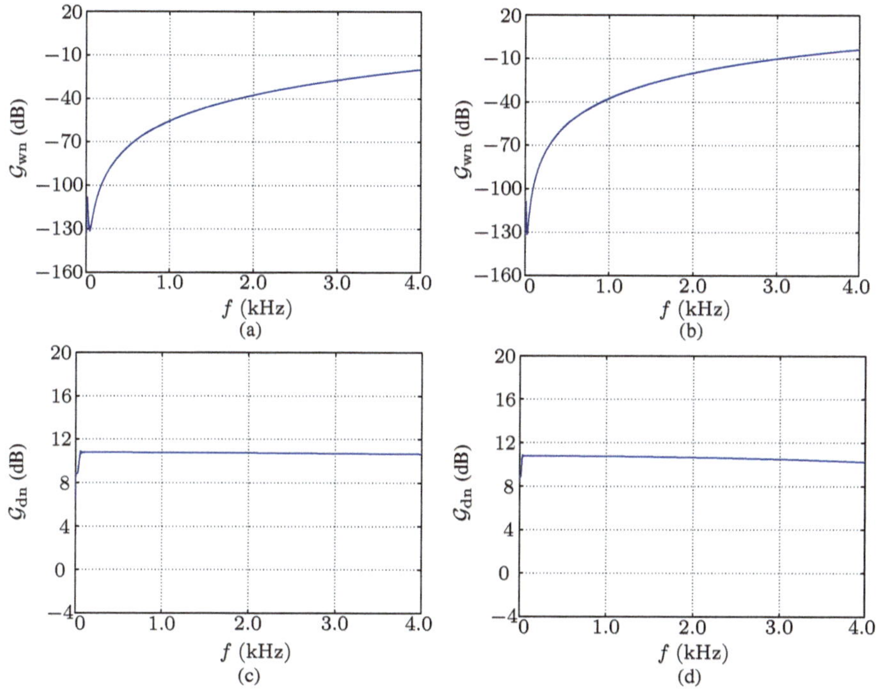


**Fig. 6.10** The white noise gain and directivity factor of the superdirective beamformer with a UCA ($M = 6$ and $\theta_s = 0°$), with the symmetry constraint, as a function of frequency and two values of $\delta$: (a) the white noise gain for $\delta = 1$ cm, (b) the white noise gain for $\delta = 2$ cm, (c) the directivity factor for $\delta = 1$ cm, and (d) the directivity factor for $\delta = 2$ cm.

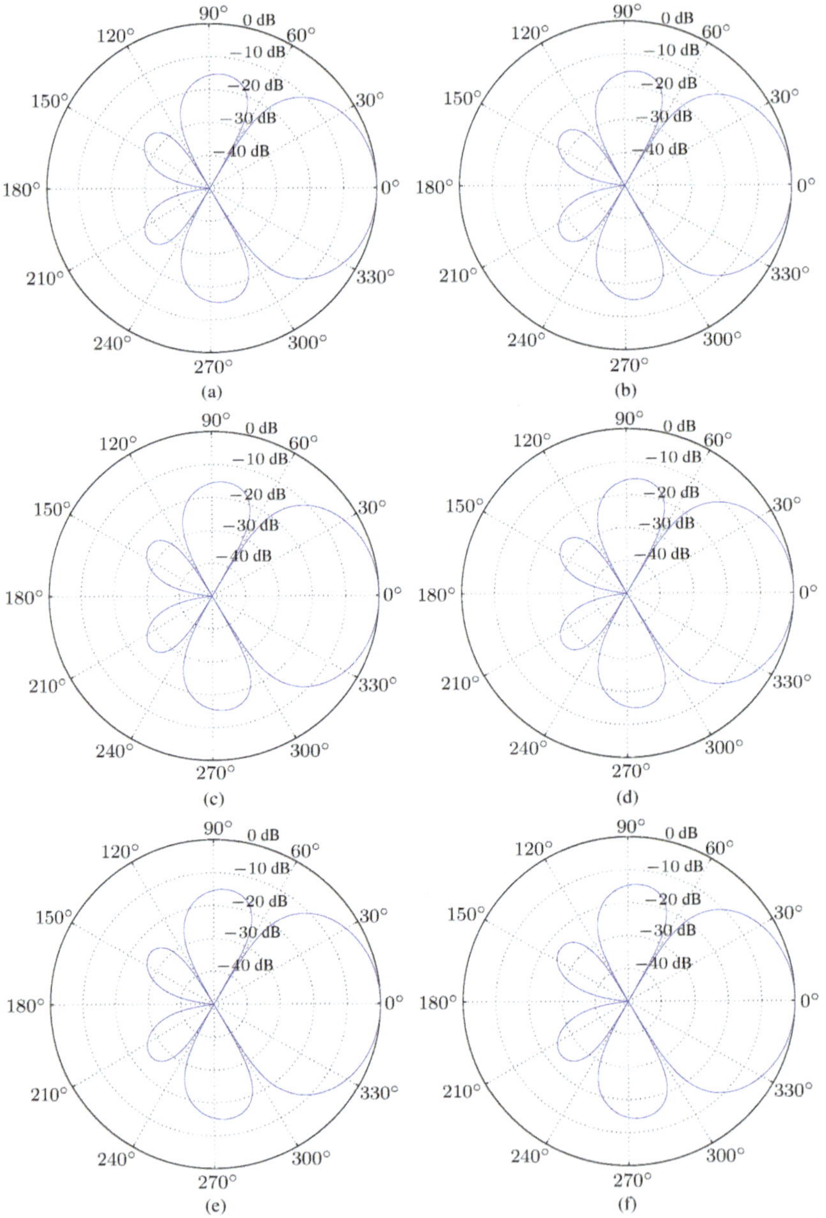


**Fig. 6.11** Patterns of the superdirective beamformer with a UCA ($M = 6$ and $\theta_s = 0°$), with the maximum of nulls, for several frequencies and two values of $\delta$: (a) $f = 0.5$ kHz, $\delta = 1$ cm, (b) $f = 0.5$ kHz, $\delta = 2$ cm, (c) $f = 1$ kHz, $\delta = 1$ cm, (d) $f = 1$ kHz, $\delta = 2$ cm, (e) $f = 3$ kHz, $\delta = 1$ cm, and (f) $f = 3$ kHz, $\delta = 2$ cm.

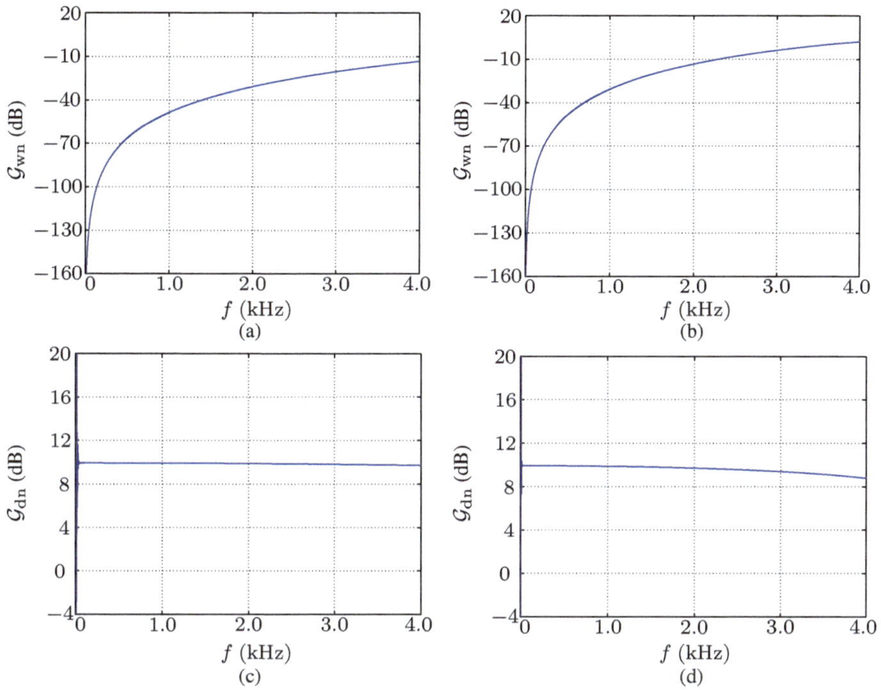


**Fig. 6.12** The white noise gain and directivity factor of the superdirective beamformer with a UCA ($M = 6$ and $\theta_s = 0°$), with the maximum of nulls, as a function of frequency and two values of $\delta$: (a) the white noise gain for $\delta = 1$ cm, (b) the white noise gain for $\delta = 2$ cm, (c) the directivity factor for $\delta = 1$ cm, and (d) the directivity factor for $\delta = 2$ cm.

# References


1. H. Cox, R. M. Zeskind, and T. Kooij, "Practical supergain," *IEEE Trans. Acoust., Speech, Signal Process.*, vol. ASSP-34, pp. 393–398, June 1986.
2. A. I. Uzkov, "An approach to the problem of optimum directive antenna design," *Comptes Rendus (Doklady) de l'Academie des Sciences de l'URSS*, vol. LIII, no. 1, pp. 35–38, 1946.
3. J. Benesty and J. Chen, *Study and Design of Differential Microphone Arrays.* Berlin, Germany: Springer-Verlag, 2012.
4. H. Cox, R. M. Zeskind, and M. M. Owen, "Robust adaptive beamforming," *IEEE Trans. Acoust., Speech, Signal Process.*, vol. ASSP-35, pp. 1365–1376, Oct. 1987.
5. S. Yan and Y. Ma, "Robust supergain beamforming for circular array via second-order cone programming," *Applied Acoustics*, vol. 66, pp. 1018–1032, 2005.
6. P. J. Davis, *Circulant Matrices.* New York, USA: John Wiley & Sons, Inc., 1979.
7. C. Pan, J. Chen, and J. Benesty, "Performance study of the MVDR beamformer as a function of the source incident angle," *IEEE/ACM Trans. Audio, Speech, Lang. Process.*, vol. 22, pp. 67–79, Jan. 2014.

# Chapter 7
# Minimum-Norm Solution for Robust Circular Differential Arrays

In this chapter, we show how the classical adaptive beamforming technique is related to differential circular arrays. From this important and useful relationship, we then derive a minimum-norm filter for the design of any order differential array, which can be robust against white noise amplification. This approach exploits the fact that the number of microphones can be much larger than the order of the CDMA. As a result, the more microphones, the more robust is the CDMA for a predetermined order.

## 7.1 Adaptive Beamforming versus Circular Differential Arrays

A very important question one may ask: how is adaptive beamforming [1] related to (circular) differential arrays [2], [3]?

We know from the previous chapters that any CDMA of order $N = \lfloor \frac{M}{2} \rfloor$ can be designed by solving the linear system of $N' = N + 1 + M'$ equations:

$$\mathbf{A}(\omega, \theta) \mathbf{h}(\omega) = \mathbf{b}, \tag{7.1}$$

where

$$M' = M - \lfloor \frac{M}{2} \rfloor - 1 \tag{7.2}$$

is the number of constraints for the symmetry of the coefficients of the filter, $M$ is the number of microphones,

$$\mathbf{A}(\omega, \theta) = \begin{bmatrix} \mathbf{D}(\omega, \boldsymbol{\theta}) \\ \mathbf{C} \end{bmatrix} \tag{7.3}$$

is the constraint matrix of size $N' \times M$,



J. Benesty et al., *Design of Circular Differential Microphone Arrays*,
Springer Topics in Signal Processing 12, DOI 10.1007/978-3-319-14842-7_7

$$\mathbf{D}(\omega,\boldsymbol{\theta}) = \begin{bmatrix} \mathbf{d}^H(\omega,0) \\ \mathbf{d}^H(\omega,\theta_{N,1}) \\ \vdots \\ \mathbf{d}^H(\omega,\theta_{N,N}) \end{bmatrix} \tag{7.4}$$

is a matrix of size $(N+1) \times M$, $\mathbf{d}(\omega,\theta_{N,n})$ is the steering vector of length $M$,

$$\mathbf{C} = \begin{bmatrix} \mathbf{c}_{M,1}^T \\ \mathbf{c}_{M,2}^T \\ \vdots \\ \mathbf{c}_{M,M'}^T \end{bmatrix} \tag{7.5}$$

is a matrix of size $M' \times M$, with the vector $\mathbf{c}_{M,m'}$ being defined in Chapter 6,

$$\mathbf{h}(\omega) = \begin{bmatrix} H_1(\omega) \; H_2(\omega) \; \cdots \; H_M(\omega) \end{bmatrix}^T \tag{7.6}$$

is a filter of length $M$,

$$\mathbf{b} = \begin{bmatrix} \boldsymbol{\beta}^T \; 0 \cdots 0 \end{bmatrix}^T \tag{7.7}$$

is a vector of length $N'$, and

$$\boldsymbol{\theta} = \begin{bmatrix} 0 \; \theta_{N,1} \; \cdots \; \theta_{N,N} \end{bmatrix}^T \tag{7.8}$$

$$\boldsymbol{\beta} = \begin{bmatrix} 1 \; \beta_{N,1} \; \cdots \; \beta_{N,N} \end{bmatrix}^T \tag{7.9}$$

are vectors of length $N+1$ containing the design coefficients of the directivity pattern. In all previous chapters, the case $N = \lfloor \frac{M}{2} \rfloor$ (or, equivalently, $M = N'$) was considered but we may want to choose $\lfloor \frac{M}{2} \rfloor$ (or $M$) much larger than $N$ (or $N'$).

In adaptive beamforming, we minimize the residual noise at the beamformer output subject to some constraints. Here, the constraints are summarized in (7.1). Mathematically, this procedure is equivalent to

$$\min_{\mathbf{h}(\omega)} \mathbf{h}^H(\omega)\,\boldsymbol{\Phi}_{\mathbf{v}}(\omega)\,\mathbf{h}(\omega) \quad \text{subject to} \quad \mathbf{A}(\omega,\theta)\,\mathbf{h}(\omega) = \mathbf{b}, \tag{7.10}$$

for which the solution is

$$\mathbf{h}_{\text{LCMV}}(\omega) = \boldsymbol{\Phi}_{\mathbf{v}}^{-1}(\omega)\,\mathbf{A}^H(\omega,\theta)\left[\mathbf{A}(\omega,\theta)\,\boldsymbol{\Phi}_{\mathbf{v}}^{-1}(\omega)\,\mathbf{A}^H(\omega,\theta)\right]^{-1}\mathbf{b}, \tag{7.11}$$

where we recognize the well-known linearly constrained minimum variance (LCMV) filter [4], [5], [6]. We observe that for the matrix $\mathbf{A}(\omega,\theta)\,\boldsymbol{\Phi}_{\mathbf{v}}^{-1}(\omega)\,\mathbf{A}^H(\omega,\theta)$ in (7.11) to be full rank, we must have $N' \leq M$, i.e., $N \leq \lfloor \frac{M}{2} \rfloor$, which is the same condition to design a circular differential array of order $N$. It can also be shown that (7.11) can be expressed as

$$\mathbf{h}_{\mathrm{LCMV}}(\omega) = \mathbf{\Phi}_{\mathbf{y}}^{-1}(\omega)\mathbf{A}^{H}(\omega,\theta)\left[\mathbf{A}(\omega,\theta)\mathbf{\Phi}_{\mathbf{y}}^{-1}(\omega)\mathbf{A}^{H}(\omega,\theta)\right]^{-1}\mathbf{b}, \quad (7.12)$$

where $\mathbf{\Phi}_{\mathbf{y}}(\omega) = E\left[\mathbf{y}(\omega)\mathbf{y}^{H}(\omega)\right]$ is the correlation matrix of the observation signal vector, $\mathbf{y}(\omega)$. However, from an implementation point of view, it is preferable to use (7.11) than (7.12) since $\mathbf{\Phi}_{\mathbf{v}}(\omega)$ is usually better conditioned than $\mathbf{\Phi}_{\mathbf{y}}(\omega)$.

For $M = N'$, we easily deduce from (7.11) [or (7.12)] that

$$\mathbf{h}_{\mathrm{LCMV}}(\omega) = \mathbf{A}^{-1}(\omega,\theta)\mathbf{b}, \quad (7.13)$$

which corresponds exactly to the filter of an $\lfloor \frac{M}{2} \rfloor$th-order CDMA or the solution of (7.1), which is a linear system of $M$ equations and $M$ unknowns.

For $M > N'$ and spatially white noise, (7.11) becomes

$$\mathbf{h}_{\mathrm{LCMV}}(\omega) = \mathbf{A}^{H}(\omega,\theta)\left[\mathbf{A}(\omega,\theta)\mathbf{A}^{H}(\omega,\theta)\right]^{-1}\mathbf{b}, \quad (7.14)$$

which corresponds to the minimum-norm solution of (7.1). This shows that the LCMV filter is fundamentally related to the filter of an $\lfloor \frac{M}{2} \rfloor$th-order CDMA.

Even though we minimize the residual noise at the beamformer output, there is no guaranty with the LCMV filter that the white noise gain should be greater than 1; this is due to the multiple constraints that need to be fulfilled at the same time. However, by virtue of the inclusion principle [7], the solution for $M > N'$ is better than the solution for $M = N'$ when it comes to white noise amplification. In fact, the greater is $M$, the better is the solution. Therefore, we can build a high-order CDMA without amplifying much the white noise by having many more microphones as compared to the order of the CDMA.

We can continue our comparisons to the maximum SNR filter without the symmetry constraint:

$$\mathbf{h}_{\max}(\omega) = \frac{\mathbf{\Phi}_{\mathbf{v}}^{-1}(\omega)\mathbf{d}(\omega,0)}{\mathbf{d}^{H}(\omega,0)\mathbf{\Phi}_{\mathbf{v}}^{-1}(\omega)\mathbf{d}(\omega,0)} \quad (7.15)$$

or, equivalently,

$$\mathbf{h}_{\max}(\omega) = \frac{\mathbf{\Phi}_{\mathbf{y}}^{-1}(\omega)\mathbf{d}(\omega,0)}{\mathbf{d}^{H}(\omega,0)\mathbf{\Phi}_{\mathbf{y}}^{-1}(\omega)\mathbf{d}(\omega,0)}. \quad (7.16)$$

This filter reduces more noise than the LCMV filter and the gain in SNR is always greater than or equal to 1 (for all types of noise) [8] but $\mathbf{\Phi}_{\mathbf{v}}(\omega)$ or $\mathbf{\Phi}_{\mathbf{y}}(\omega)$ needs to be estimated accurately; otherwise, the estimation errors may lead to some desired signal cancellation. Moreover, the pattern of this maximum SNR filter does not correspond, in general, to any known directivity pattern contrary to the LCMV filter. Replacing $\mathbf{\Phi}_{\mathbf{v}}(\omega)$ in (7.15) by $\mathbf{\Gamma}_{\mathrm{dn}}(\omega)$, we get the maximum SNR filter derived in Chapter 6.

With the symmetry constraint, it is not hard to show that the maximum SNR filter is

$$\mathbf{h}_{\mathrm{max}}(\omega) = \mathbf{\Phi}_{\mathbf{v}}^{-1}(\omega)\mathbf{C}^{H}(\omega)\left[\mathbf{C}(\omega)\mathbf{\Phi}_{\mathbf{v}}^{-1}(\omega)\mathbf{C}^{H}(\omega)\right]^{-1}\mathbf{i}_{1}, \tag{7.17}$$

or, equivalently,

$$\mathbf{h}_{\mathrm{max}}(\omega) = \mathbf{\Phi}_{\mathbf{y}}^{-1}(\omega)\mathbf{C}^{H}(\omega)\left[\mathbf{C}(\omega)\mathbf{\Phi}_{\mathbf{y}}^{-1}(\omega)\mathbf{C}^{H}(\omega)\right]^{-1}\mathbf{i}_{1}, \tag{7.18}$$

where $\mathbf{C}(\omega)$ and $\mathbf{i}_1$ are defined in Chapter 6. Again, this filter reduces more noise than the LCMV filter (but less than the maximum SNR filter without the symmetry constraint) and the gain in SNR is always greater than or equal to 1 (for all types of noise). When $\mathbf{\Phi}_{\mathbf{v}}(\omega)$ in (7.17) is replaced by $\mathbf{\Gamma}_{\mathrm{dn}}(\omega)$, we get the pattern of the hypercardioid (of order $\lfloor \frac{M}{2} \rfloor$) but the white noise can be highly amplified since when the number of microphones is increased so is the order of the hypercardioid. This optimization procedure maximizes the directivity factor at the expense of white noise amplification. While in the LCMV filter, we can increase the number of microphones without increasing much the order of the CDMA; as a result, white noise amplification can be controlled. We will get back to this point in the next section.

## 7.2 Minimum-Norm Filter

As it is clearly explained in the previous section, the minimum-norm solution given in (7.14) is of great importance here since it has the potential to solve the problem of white noise amplification encountered in all known differential arrays. Let us slightly change the notation of this filter and denote it by

$$\mathbf{h}(\omega, \boldsymbol{\theta}, \boldsymbol{\beta}) = \mathbf{A}^{H}(\omega, \theta)\left[\mathbf{A}(\omega, \theta)\mathbf{A}^{H}(\omega, \theta)\right]^{-1}\mathbf{b}, \tag{7.19}$$

where, again, the vectors $\boldsymbol{\theta}$ and $\boldsymbol{\beta}$ of length $N+1$ determine the pattern and the order of the CDMA. Basically, the length of these vectors determines (roughly) the order of the CDMA while their values determine the pattern. Meanwhile, the length, $M$, of the minimum-norm filter, $\mathbf{h}(\omega, \boldsymbol{\theta}, \boldsymbol{\beta})$, can be much larger than $N'$, which will help make the CDMA robust against white noise amplification. In our context, a more rigorous way to derive the minimum-norm filter is by maximizing the white noise gain subject to the $N$th-order CDMA fundamental constraints. This is equivalent to minimizing $\mathbf{h}^{H}(\omega, \boldsymbol{\theta}, \boldsymbol{\beta})\mathbf{h}(\omega, \boldsymbol{\theta}, \boldsymbol{\beta})$ subject to (7.1), from which (7.19) results. In other words, the LCMV filter in the presence of white noise is the solution we are looking for. Therefore, for a large number of microphones, the white noise gain should approach $M$. If the value of $M$ is much larger than $N'$, the order of the CDMA may not be equal to $N$ anymore but the $N$th-order CDMA fundamental constraints will always be fulfilled. As a result, the resulting shape of the directivity pattern may slightly be different than the one obtained with $M = N'$. This approach is the best we can do to solve the conflicting requirement of a high-order CDMA which does not amplify the white noise.

It is easy to see that the beampattern of the minimum-norm filter is

$$\begin{aligned}\mathcal{B}\left[\mathbf{h}\left(\omega,\boldsymbol{\theta},\boldsymbol{\beta}\right),\theta\right] &= \mathbf{h}^H\left(\omega,\boldsymbol{\theta},\boldsymbol{\beta}\right)\mathbf{d}\left(\omega,\theta\right) \\ &= \mathbf{b}^T\left[\mathbf{A}\left(\omega,\theta\right)\mathbf{A}^H\left(\omega,\theta\right)\right]^{-1}\mathbf{A}\left(\omega,\theta\right)\mathbf{d}\left(\omega,\theta\right).\end{aligned} \tag{7.20}$$

We also deduce that the white noise gain, the directivity factor, and the gain for a point noise source are, respectively,

$$\mathcal{G}_{\text{wn}}\left[\mathbf{h}\left(\omega,\boldsymbol{\theta},\boldsymbol{\beta}\right)\right] = \frac{1}{\mathbf{b}^T\left[\mathbf{A}\left(\omega,\theta\right)\mathbf{A}^H\left(\omega,\theta\right)\right]^{-1}\mathbf{b}}, \tag{7.21}$$

$$\mathcal{G}_{\text{dn}}\left[\mathbf{h}\left(\omega,\boldsymbol{\theta},\boldsymbol{\beta}\right)\right] = \frac{1}{\mathbf{h}^H\left(\omega,\boldsymbol{\theta},\boldsymbol{\beta}\right)\boldsymbol{\Gamma}_{\text{dn}}\left(\omega\right)\mathbf{h}\left(\omega,\boldsymbol{\theta},\boldsymbol{\beta}\right)}, \tag{7.22}$$

and

$$\mathcal{G}_{\text{ns}}\left[\mathbf{h}\left(\omega,\boldsymbol{\theta},\boldsymbol{\beta}\right)\right] = \frac{1}{\left|\mathcal{B}\left[\mathbf{h}\left(\omega,\boldsymbol{\theta},\boldsymbol{\beta}\right),\theta_{\text{n}}\right]\right|^2}, \tag{7.23}$$

where $\theta_{\text{n}}$ is the angle of the point noise source (see Chapter 2).

Ideally, we would like to have

$$\mathcal{G}_{\text{dn}}\left[\mathbf{h}\left(\omega,\boldsymbol{\theta},\boldsymbol{\beta}\right)\right] = \mathcal{G}_N, \tag{7.24}$$

where $\mathcal{G}_N$ is the directivity factor of the frequency-independent $N$th-order DMA (see Chapter 2) and

$$\mathcal{G}_{\text{wn}}\left[\mathbf{h}\left(\omega,\boldsymbol{\theta},\boldsymbol{\beta}\right)\right] \geq 1. \tag{7.25}$$

## 7.3 Design Examples

In this section, we show how to design several directivity patterns of different orders with the minimum-norm filter and a uniform circular array of up to eleven microphones. We only study four examples. Evidently, many more patterns can be designed with the proposed approach. In the previous chapters, we considered a bandwidth of 4 kHz; here, we use a bandwidth of 8 kHz in order to have a better insight into this new method. In all the examples, the interelement spacing is $\delta = 2$ cm for $M \leq 6$, $\delta = 1$ cm for $7 \leq M \leq 9$, and $\delta = 8$ mm for $M \geq 10$. In the rest, we choose $\theta_{\text{s}} = 0$.

### 7.3.1 First-Order Cardioid

From Chapter 3, we know that the parameters to design the first-order cardioid are summarized into the two vectors of length 2:

$$\boldsymbol{\theta} = \begin{bmatrix} 0 \ \pi \end{bmatrix}^T, \tag{7.26}$$

$$\boldsymbol{\beta} = \begin{bmatrix} 1 \ 0 \end{bmatrix}^T. \tag{7.27}$$

Figures 7.1, 7.2, and 7.3 display the patterns of the first-order cardioid with three, seven, and eleven microphones, respectively, for several frequencies. All patterns look the same, even with eleven microphones and frequencies as high as 5 kHz.

Figure 7.4 gives plots of the white noise gain of the first-order cardioid, as a function of frequency, for different values of $M$. This white noise gain increases as $M$ increases.

Figure 7.5 shows plots of the directivity factor of the first-order cardioid, as a function of frequency, for different values of $M$. This directivity factor improves as $M$ increases, especially in high frequencies.

From all simulations we have done, we can say that the minimum-norm filter is a good option to design any first-order CDMA, even with a large number of microphones.

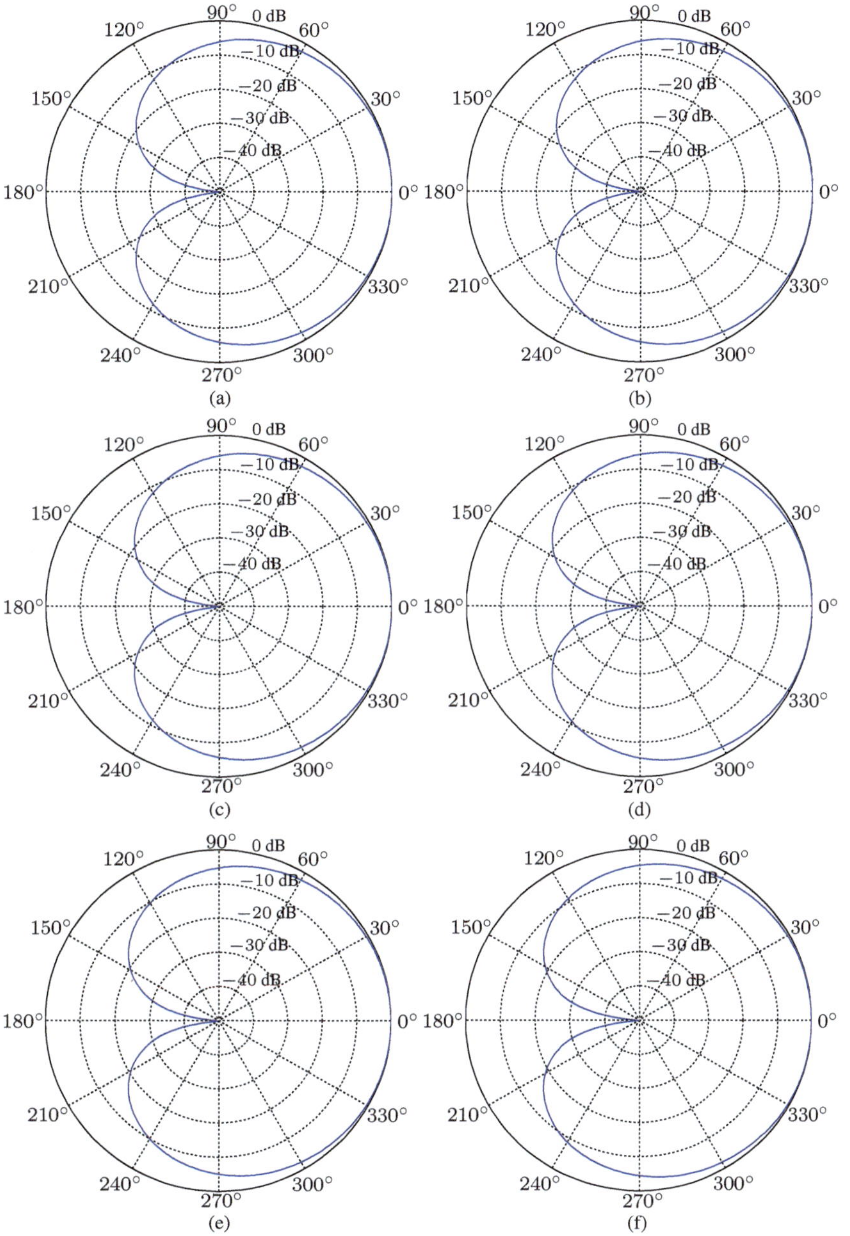


**Fig. 7.1** Patterns of the first-order cardioid with three microphones ($\delta = 2$ cm) for several frequencies: (a) $f = 0.5$ kHz, (b) $f = 1$ kHz, (c) $f = 2$ kHz, (d) $f = 3$ kHz, (e) $f = 4$ kHz, and (f) $f = 5$ kHz.

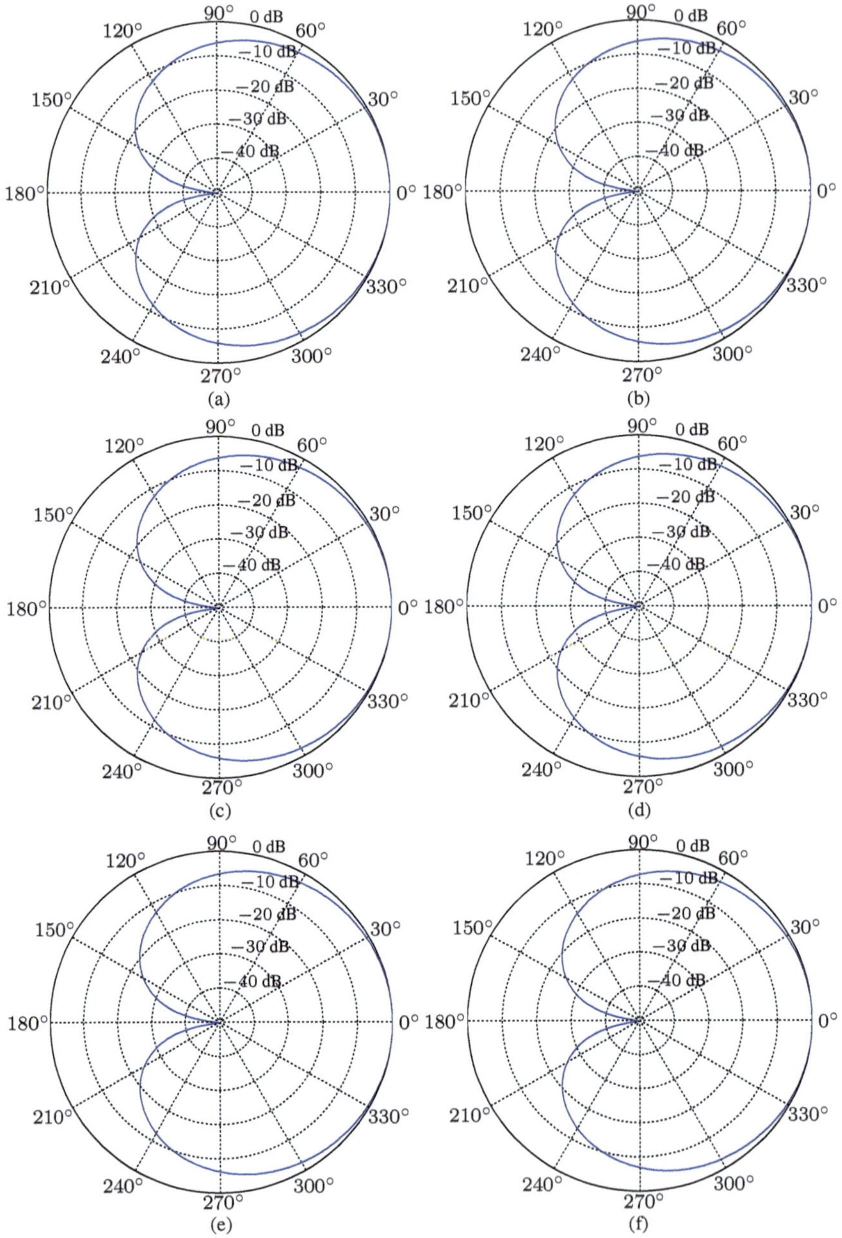


**Fig. 7.2** Patterns of the first-order cardioid with seven microphones ($\delta = 1$ cm) for several frequencies: (a) $f = 0.5$ kHz, (b) $f = 1$ kHz, (c) $f = 2$ kHz, (d) $f = 3$ kHz, (e) $f = 4$ kHz, and (f) $f = 5$ kHz.

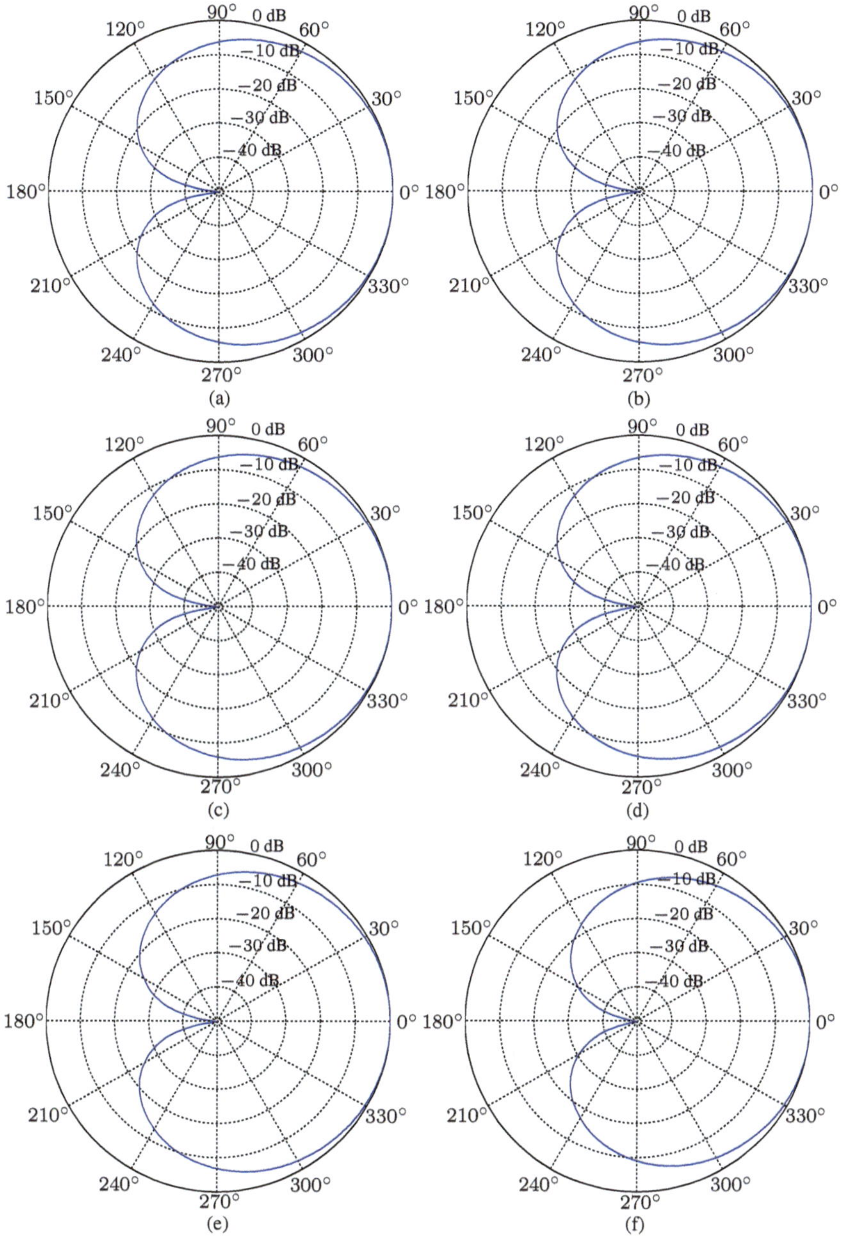


**Fig. 7.3** Patterns of the first-order cardioid with eleven microphones ($\delta = 8$ mm) for several frequencies: (a) $f = 0.5$ kHz, (b) $f = 1$ kHz, (c) $f = 2$ kHz, (d) $f = 3$ kHz, (e) $f = 4$ kHz, and (f) $f = 5$ kHz.

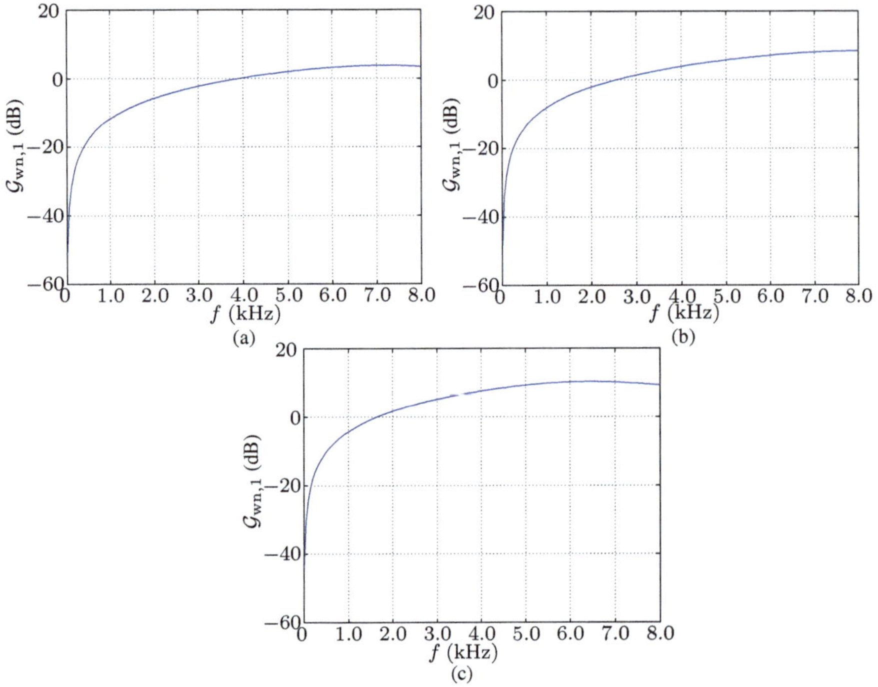


**Fig. 7.4** The white noise gain of the first-order cardioid, as a function of frequency, for different values of $M$ and $\delta$: (a) $M = 3$, $\delta = 2$ cm, (b) $M = 7$, $\delta = 1$ cm, and (c) $M = 11$, $\delta = 8$ mm.

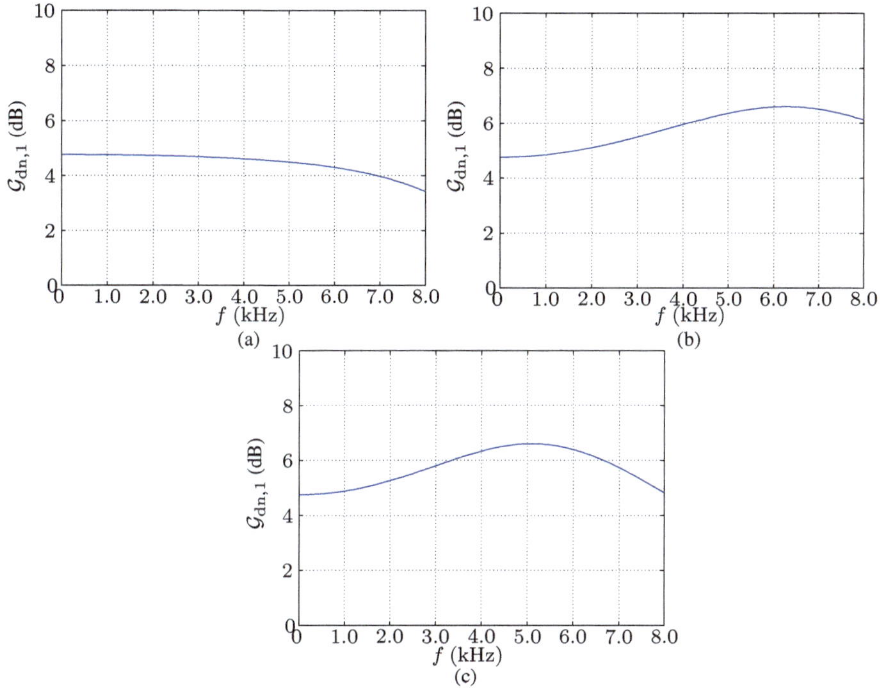


**Fig. 7.5** The directivity factor of the first-order cardioid, as a function of frequency, for different values of $M$ and $\delta$: (a) $M = 3$, $\delta = 2$ cm, (b) $M = 7$, $\delta = 1$ cm, and (c) $M = 11$, $\delta = 8$ mm.

### *7.3.2 Second-Order Cardioid*

From Chapter 4 on second-order CDMAs, we find that the parameters of the three fundamental constraints for the second-order cardioid are

$$\boldsymbol{\theta} = \left[0 \; \pi \; \frac{\pi}{2}\right]^T, \tag{7.28}$$

$$\boldsymbol{\beta} = \left[1 \; 0 \; 0\right]^T. \tag{7.29}$$

In Figs 7.6, 7.7, and 7.8, we plot the patterns of the second-order cardioid with five, nine, and eleven microphones, respectively, for several frequencies. All patterns look like the desired one, even with a large number of microphones.

In Fig. 7.9, we give plots of the white noise gain of the second-order cardioid, as a function of frequency, for different values of $M$. This white noise gain is slightly higher for larger values of $M$.

Figure 7.10 shows plots of the directivity factor of the second-order cardioid, as a function of frequency, for different values of $M$. This directivity factor is very much improved in high frequencies as $M$ is increased.

With the minimum-norm filter, we are still able to design good second-order CDMAs.

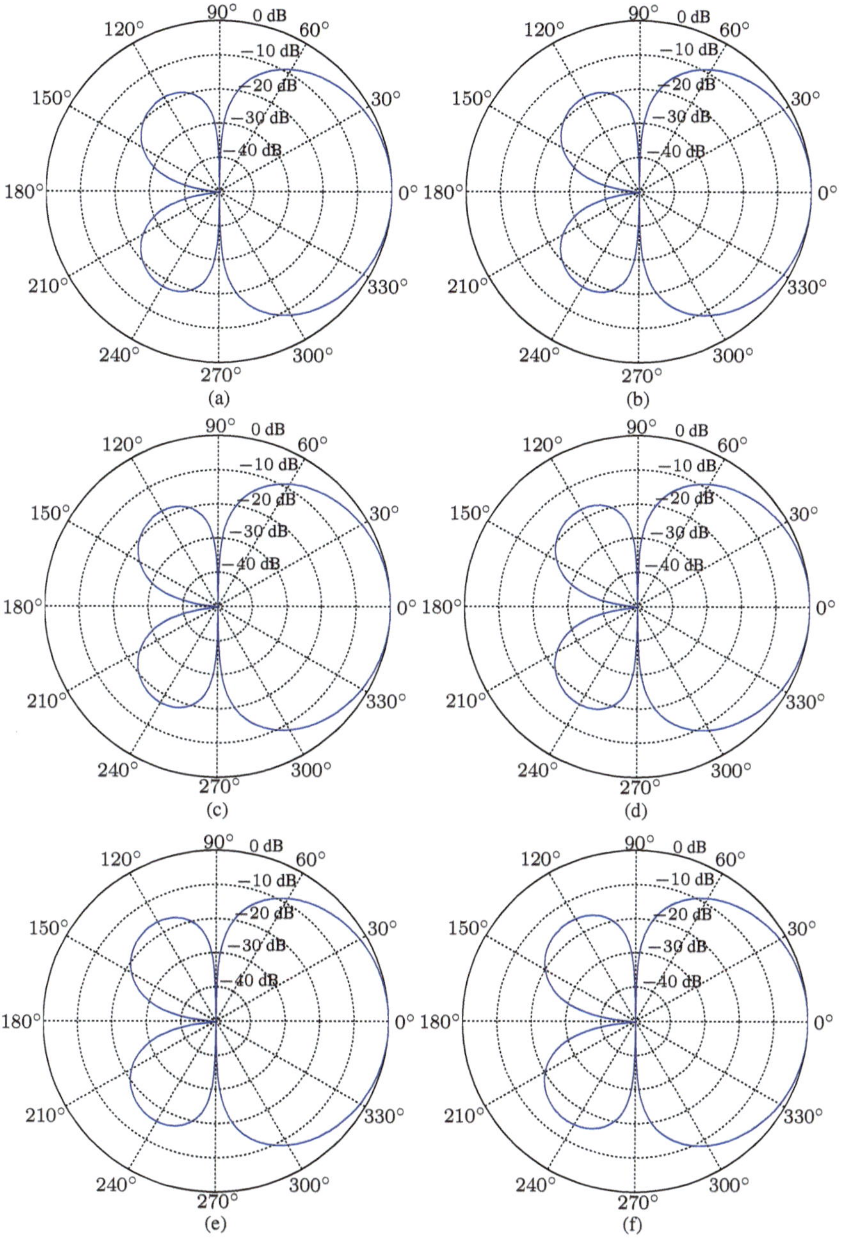


**Fig. 7.6** Patterns of the second-order cardioid with five microphones ($\delta = 2$ cm) for several frequencies: (a) $f = 0.5$ kHz, (b) $f = 1$ kHz, (c) $f = 2$ kHz, (d) $f = 3$ kHz, (e) $f = 4$ kHz, and (f) $f = 5$ kHz.

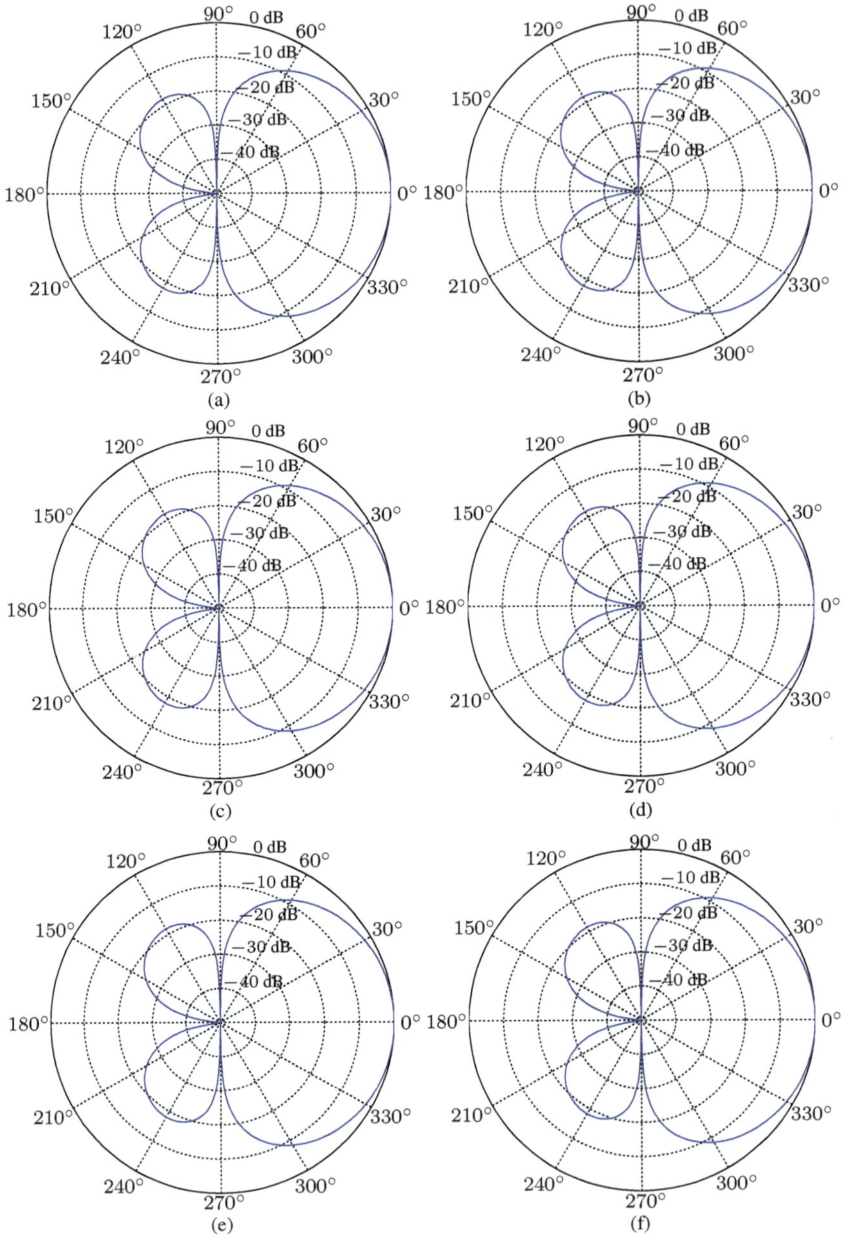


(a) (b) (c) (d) (e) (f)

**Fig. 7.7** Patterns of the second-order cardioid with nine microphones ($\delta = 1$ cm) for several frequencies: (a) $f = 0.5$ kHz, (b) $f = 1$ kHz, (c) $f = 2$ kHz, (d) $f = 3$ kHz, (e) $f = 4$ kHz, and (f) $f = 5$ kHz.

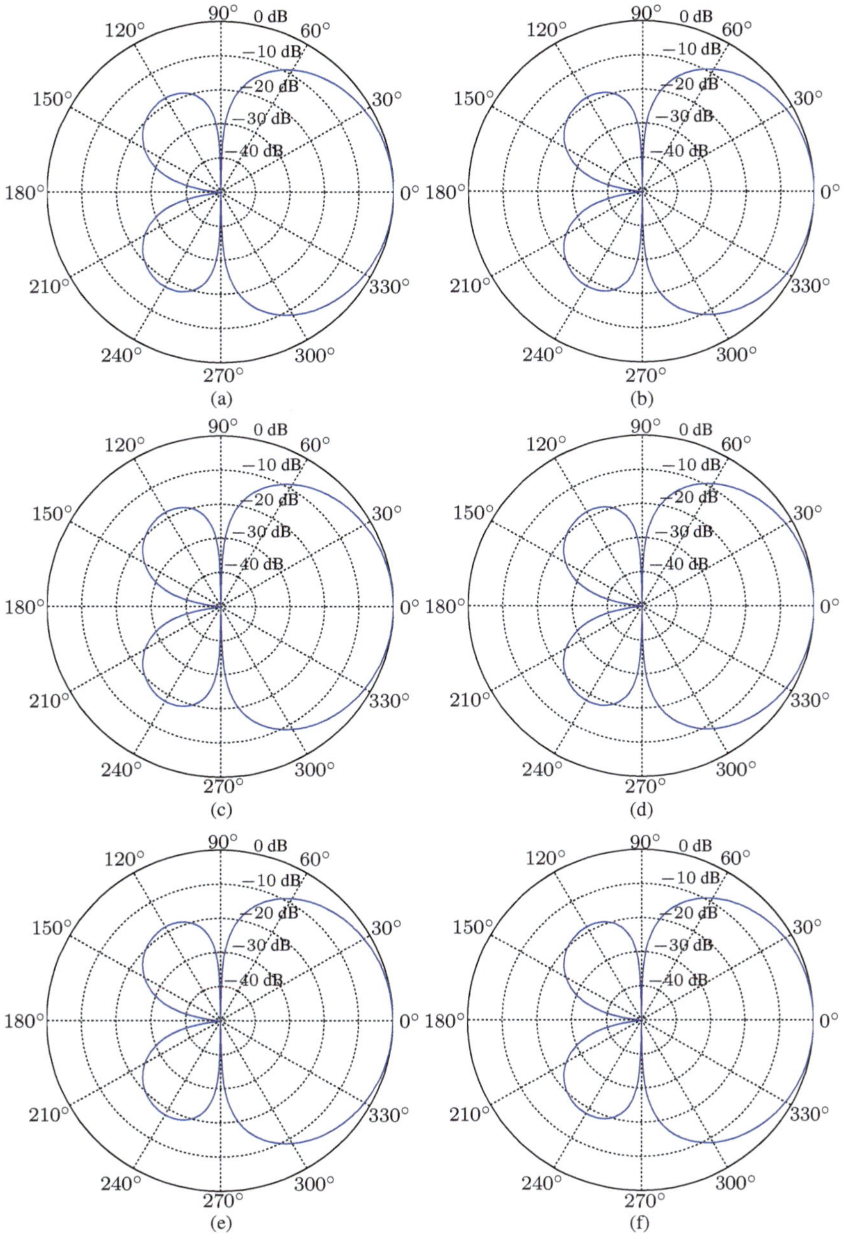


**Fig. 7.8** Patterns of the second-order cardioid with eleven microphones ($\delta = 8$ mm) for several frequencies: (a) $f = 0.5$ kHz, (b) $f = 1$ kHz, (c) $f = 2$ kHz, (d) $f = 3$ kHz, (e) $f = 4$ kHz, and (f) $f = 5$ kHz.

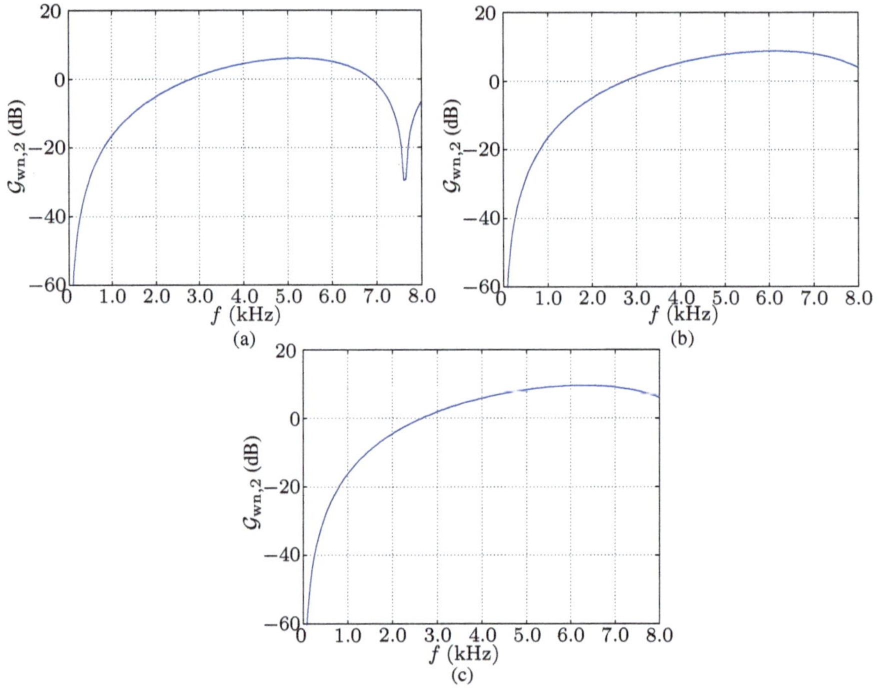


**Fig. 7.9** The white noise gain of the second-order cardioid, as a function of frequency, for different values of $M$ and $\delta$: (a) $M = 5$, $\delta = 2$ cm, (b) $M = 9$, $\delta = 1$ cm, and (c) $M = 11$, $\delta = 8$ mm.

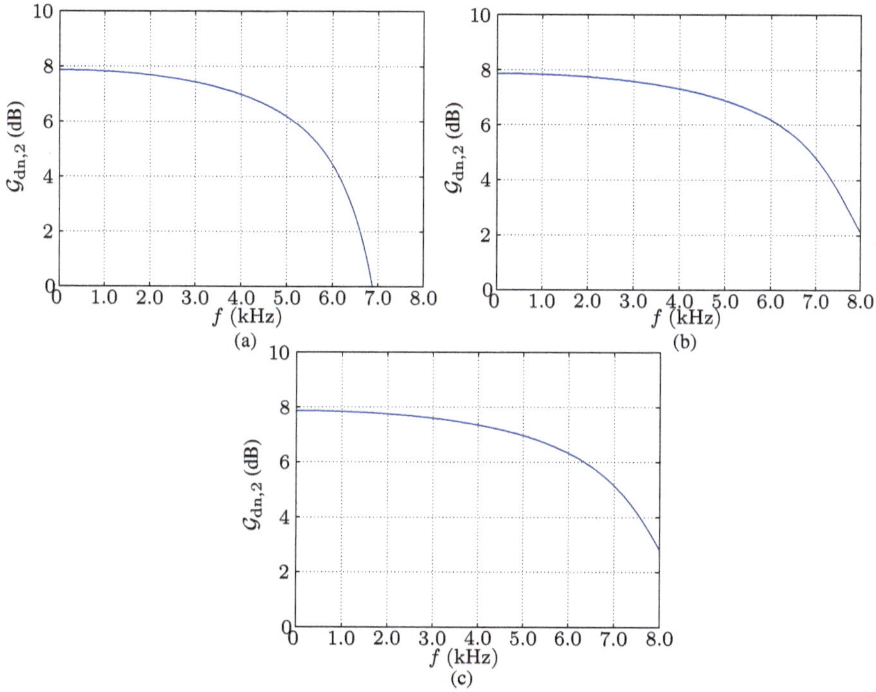


**Fig. 7.10** The directivity factor of the second-order cardioid, as a function of frequency, for different values of $M$ and $\delta$: (a) $M = 5$ and $\delta = 2$ cm, (b) $M = 9$ and $\delta = 1$ cm, and (c) $M = 11$ and $\delta = 8$ mm.

### *7.3.3 Third-Order Cardioid*

It can be checked that the parameters of the four fundamental constraints to design a third-order cardioid are summarized into the two vectors:

$$\boldsymbol{\theta} = \left[ 0 \ \pi \ \frac{\pi}{2} \ \frac{3\pi}{4} \right]^T, \tag{7.30}$$

$$\boldsymbol{\beta} = \left[ 1 \ 0 \ 0 \ -\frac{\sqrt{2}}{8} + \frac{1}{4} \right]^T. \tag{7.31}$$

Figures 7.11, 7.12, and 7.13 display the patterns of the third-order cardioid with seven, nine, and eleven microphones, respectively, for several frequencies. The designed patterns are as expected.

In Fig. 7.14, we give plots of the white noise gain of the third-order cardioid, as a function of frequency, for different values of $M$. This white noise is only very slightly improved when the number of microphones is increased.

In Fig. 7.15, we show plots of the directivity factor of the third-order cardioid, as a function of frequency, for different values of $M$. This directivity factor degrades in high frequencies when $M$ is increased.

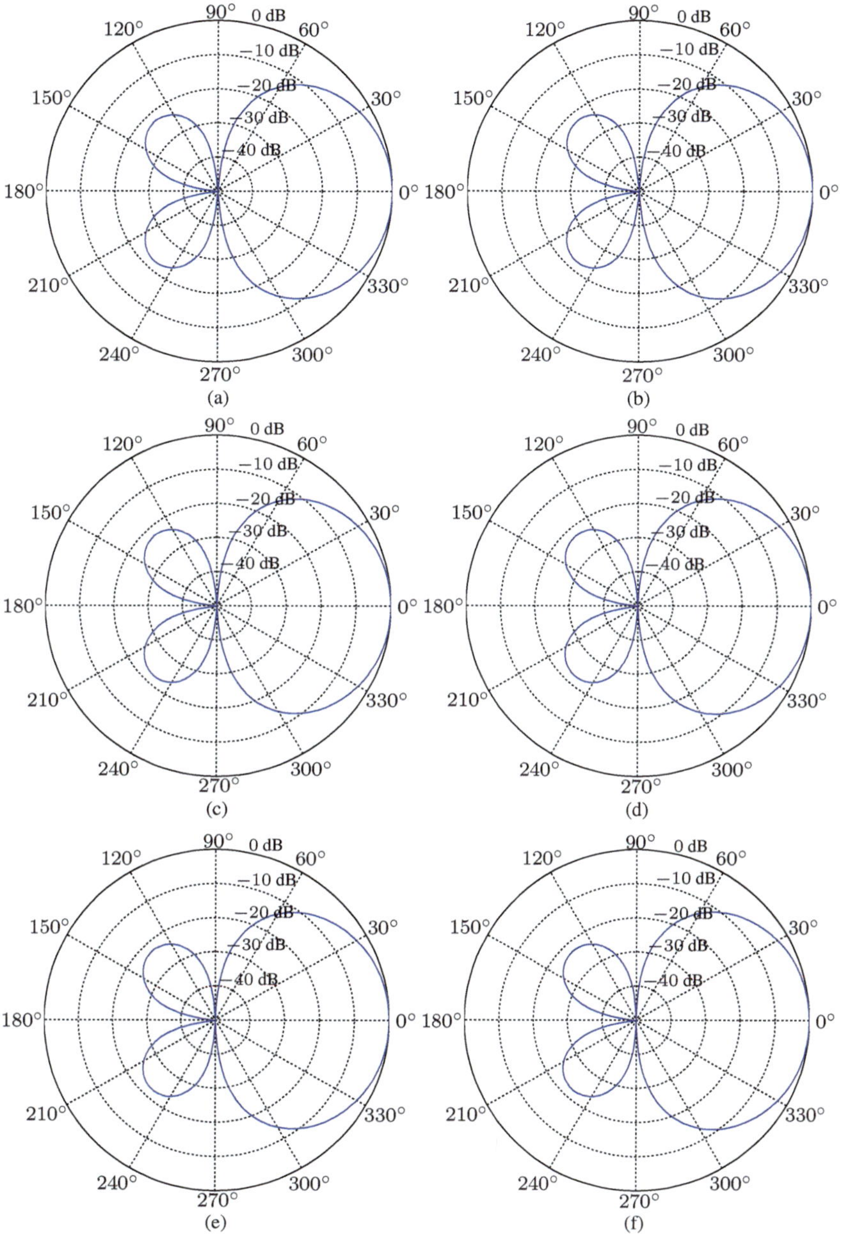


**Fig. 7.11** Patterns of the third-order cardioid with seven microphones ($\delta = 1$ cm) for several frequencies: (a) $f = 0.5$ kHz, (b) $f = 1$ kHz, (c) $f = 2$ kHz, (d) $f = 3$ kHz, (e) $f = 4$ kHz, and (f) $f = 5$ kHz.

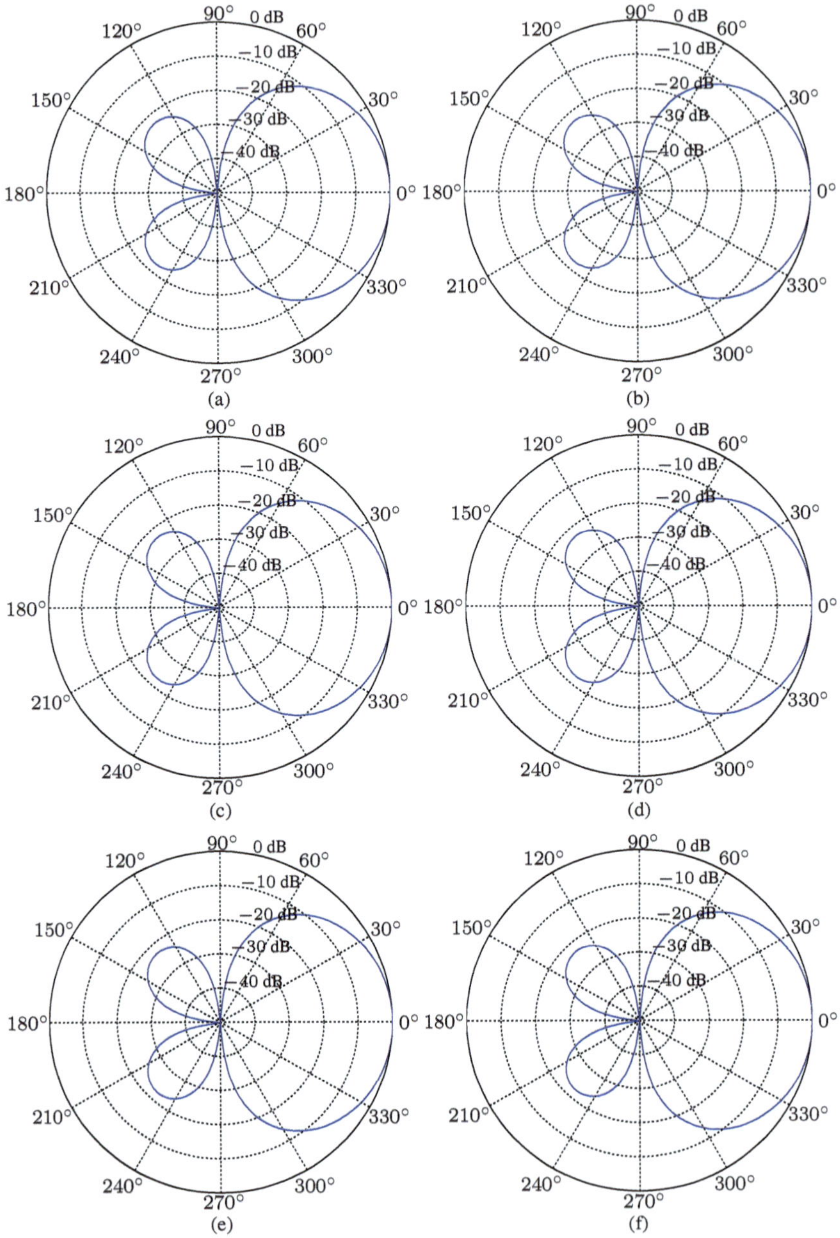


**Fig. 7.12** Patterns of the third-order cardioid with nine microphones ($\delta = 1$ cm) for several frequencies: (a) $f = 0.5$ kHz, (b) $f = 1$ kHz, (c) $f = 2$ kHz, (d) $f = 3$ kHz, (e) $f = 4$ kHz, and (f) $f = 5$ kHz.

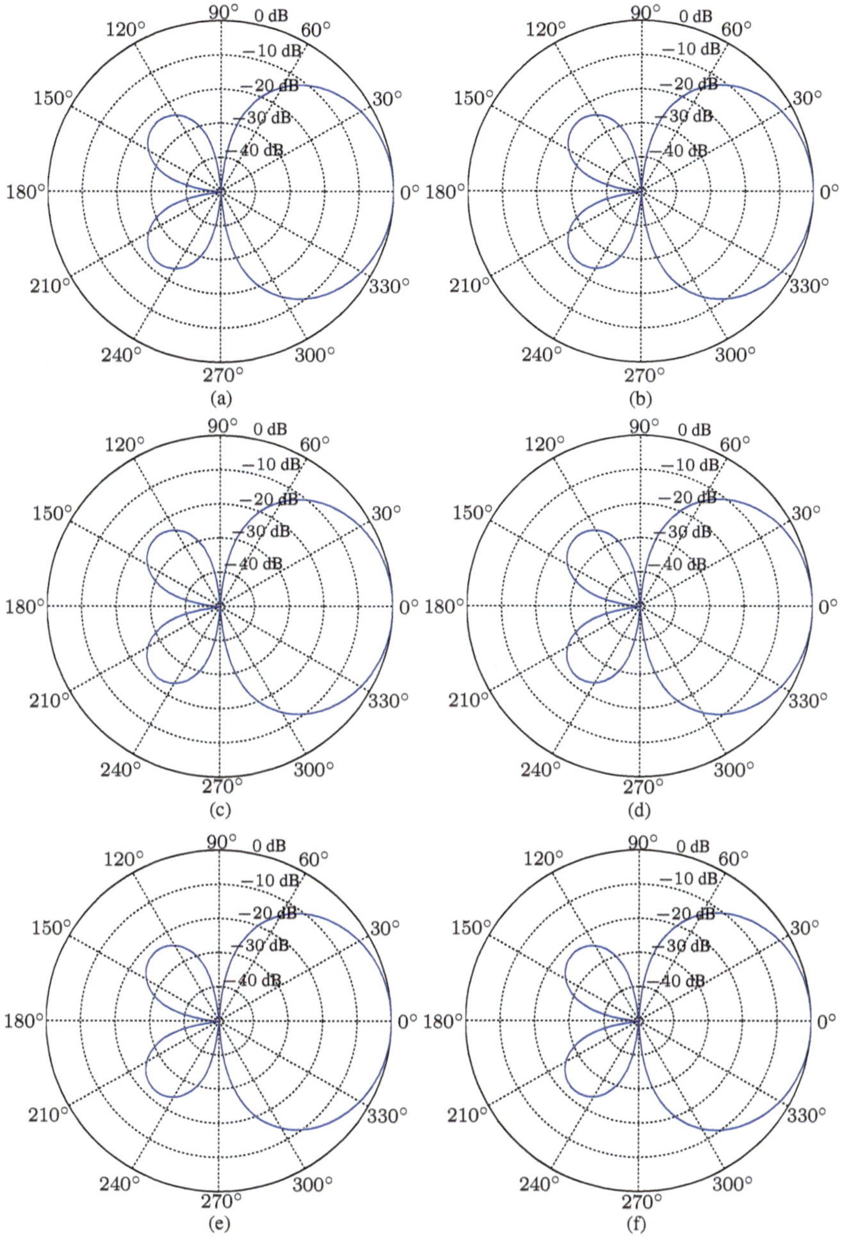


**Fig. 7.13** Patterns of the third-order cardioid with eleven microphones ($\delta = 8$ mm) for several frequencies: (a) $f = 0.5$ kHz, (b) $f = 1$ kHz, (c) $f = 2$ kHz, (d) $f = 3$ kHz, (e) $f = 4$ kHz, and (f) $f = 5$ kHz.

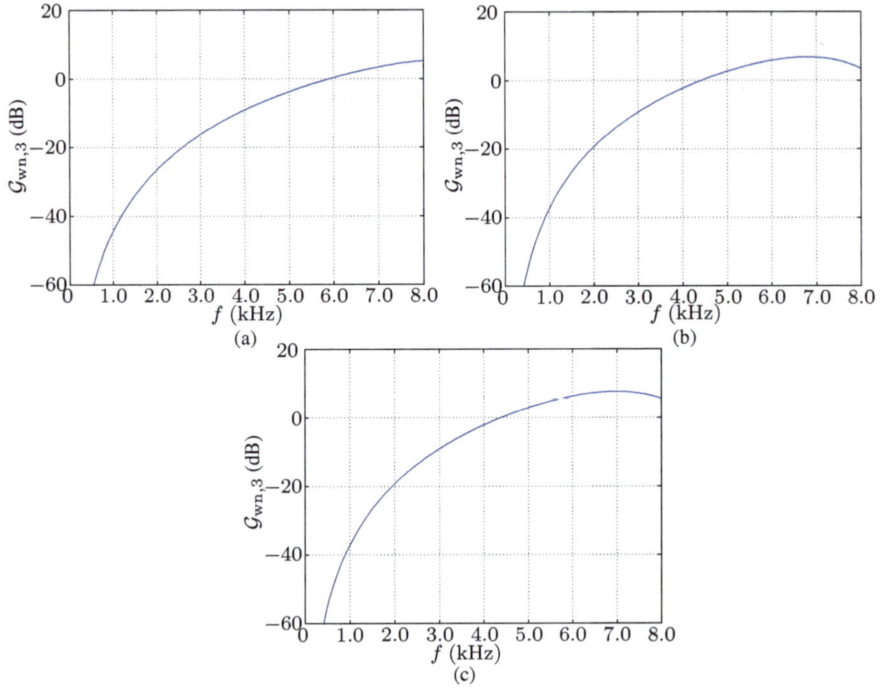


**Fig. 7.14** The white noise gain of the third-order cardioid, as a function of frequency, for different values of $M$ and $\delta$: (a) $M = 7$, $\delta = 1$ cm, (b) $M = 9$, $\delta = 1$ cm, and (c) $M = 11$, $\delta = 8$ mm.

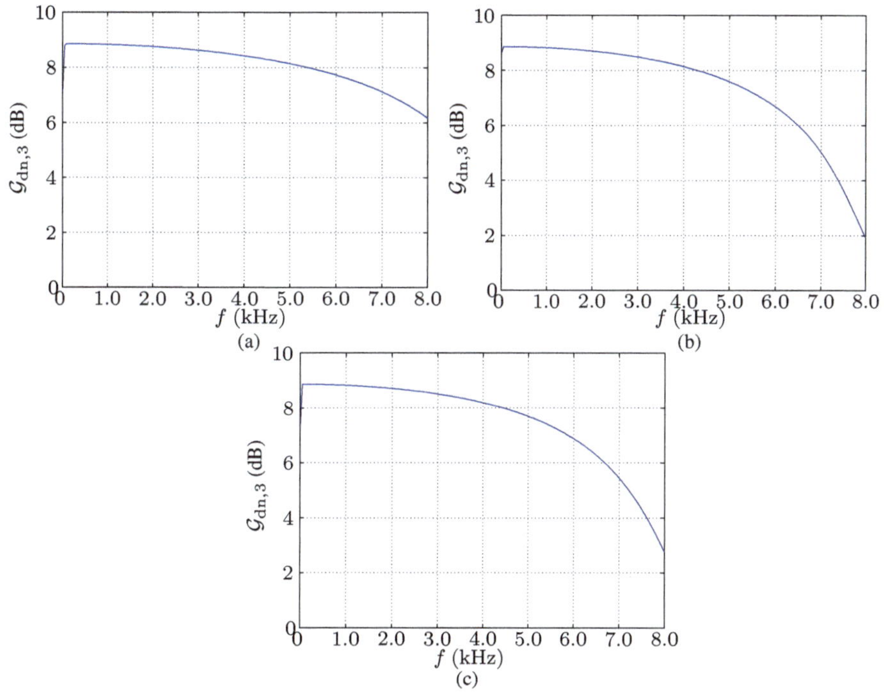


**Fig. 7.15** The directivity factor of the third-order cardioid, as a function of frequency, for different values of $M$ and $\delta$: (a) $M = 7$, $\delta = 1$ cm, (b) $M = 9$, $\delta = 1$ cm, and (c) $M = 11$, $\delta = 8$ mm.

### 7.3.4 Third-Order Pattern with Three Distinct Nulls

In this section, we study the example pattern with three distinct nulls given in Chapter 5. Therefore, we have

$$\boldsymbol{\theta} = \left[ 0 \ \frac{\pi}{2} \ \frac{2\pi}{3} \ \pi \right]^T, \tag{7.32}$$

$$\boldsymbol{\beta} = \left[ 1 \ 0 \ 0 \ 0 \right]^T. \tag{7.33}$$

Figures 7.16, 7.17, and 7.18 display the patterns of the first-order cardioid with seven, nine, and eleven microphones, respectively, for several frequencies. All patterns look very much the same.

Figure 7.19 gives plots of the white noise gain of the first-order cardioid, as a function of frequency, for different values of $M$. This white noise gain does not seem to improve much when we use more microphones.

Figure 7.20 shows plots of the directivity factor of the first-order cardioid, as a function of frequency, for different values of $M$. When we use more microphones, it seems that this directivity factor worsens in high frequencies.

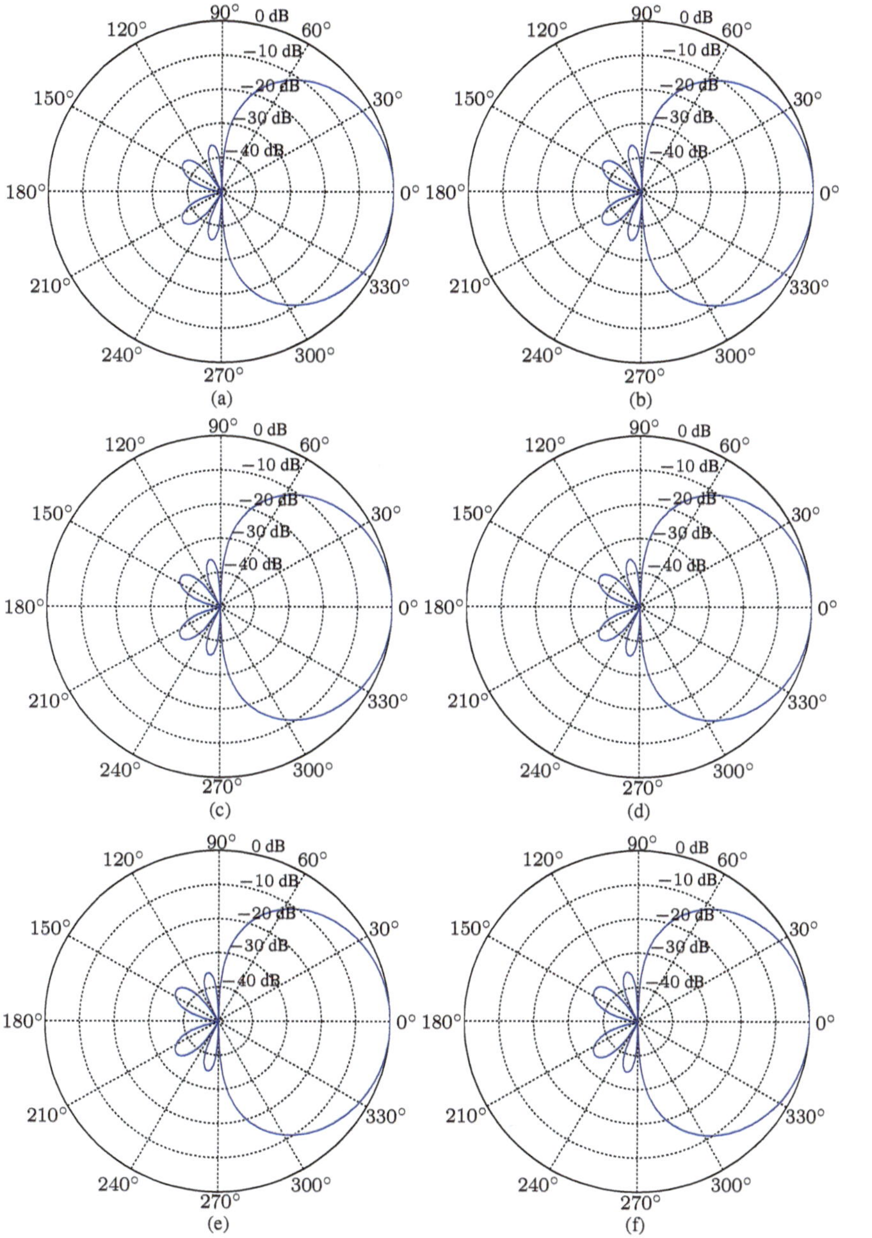


**Fig. 7.16** Patterns of a third-order CDMA with three distinct nulls and seven microphones ($\delta = 1$ cm) for several frequencies: (a) $f = 0.5$ kHz, (b) $f = 1$ kHz, (c) $f = 2$ kHz, (d) $f = 3$ kHz, (e) $f = 4$ kHz, and (f) $f = 5$ kHz.

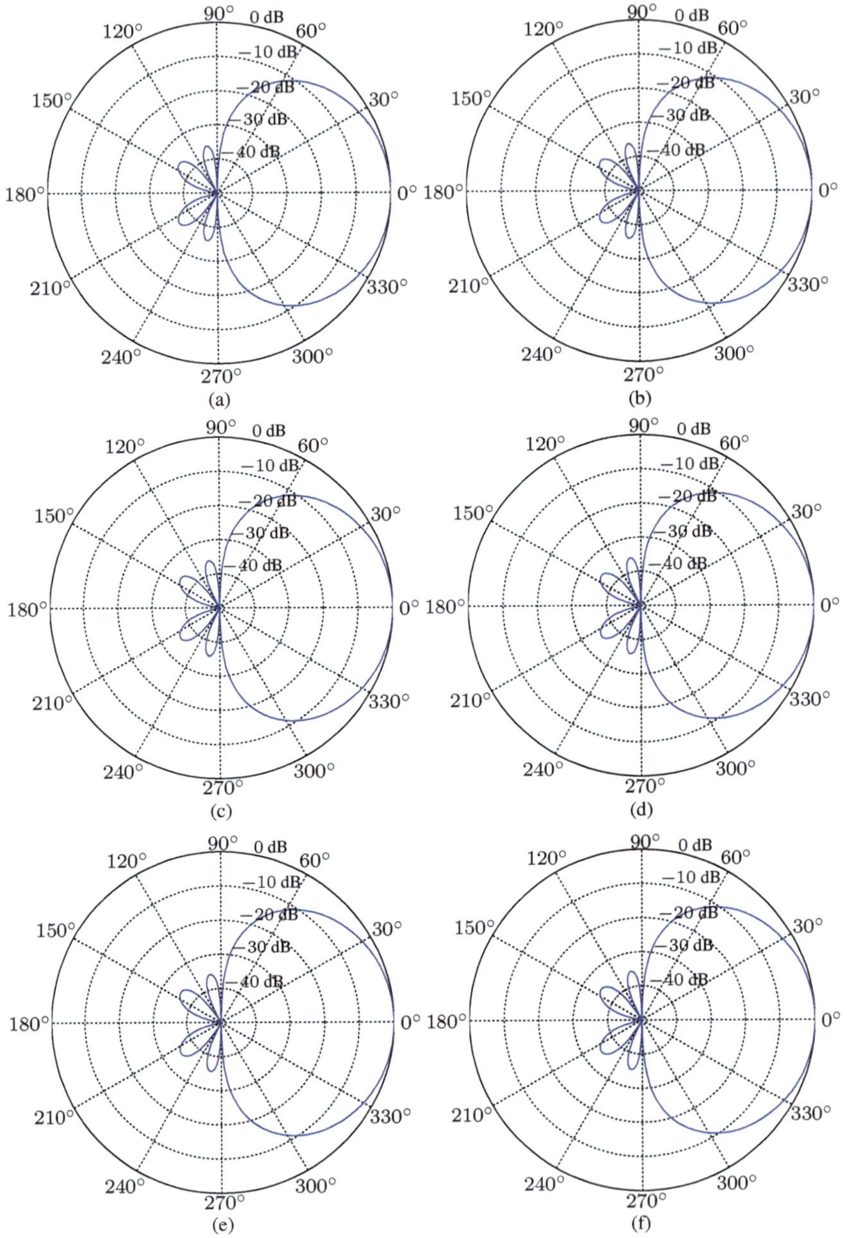


**Fig. 7.17** Patterns of a third-order CDMA with three distinct nulls and nine microphones ($\delta = 1$ cm) for several frequencies: (a) $f = 0.5$ kHz, (b) $f = 1$ kHz, (c) $f = 2$ kHz, (d) $f = 3$ kHz, (e) $f = 4$ kHz, and (f) $f = 5$ kHz.

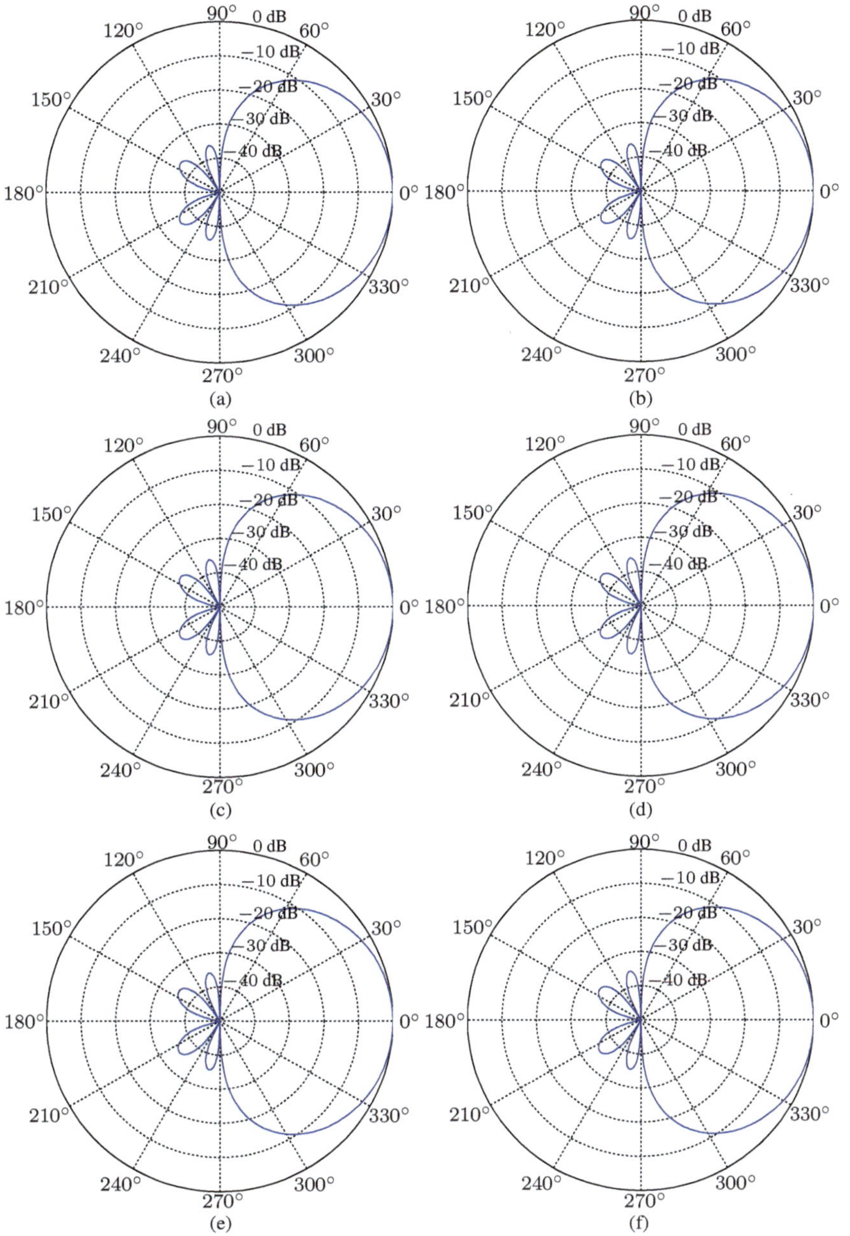


**Fig. 7.18** Patterns of a third-order CDMA with three distinct nulls and eleven microphones ($\delta = 8$ mm) for several frequencies: (a) $f = 0.5$ kHz, (b) $f = 1$ kHz, (c) $f = 2$ kHz, (d) $f = 3$ kHz, (e) $f = 4$ kHz, and (f) $f = 5$ kHz.

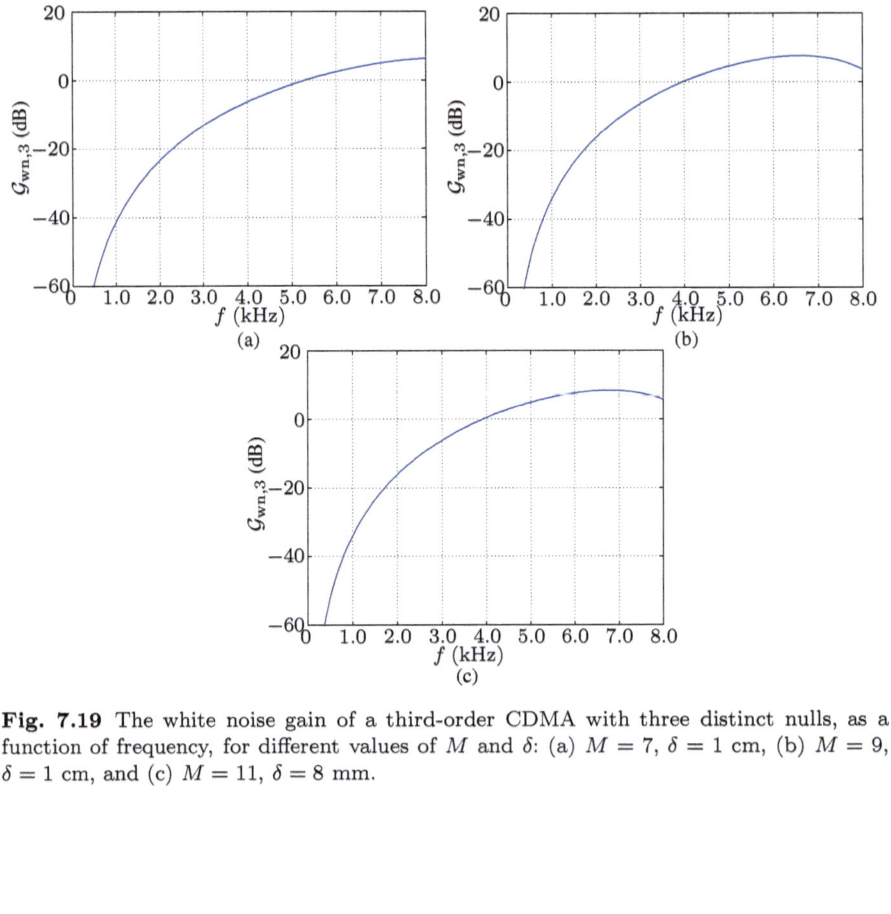


**Fig. 7.19** The white noise gain of a third-order CDMA with three distinct nulls, as a function of frequency, for different values of $M$ and $\delta$: (a) $M = 7$, $\delta = 1$ cm, (b) $M = 9$, $\delta = 1$ cm, and (c) $M = 11$, $\delta = 8$ mm.

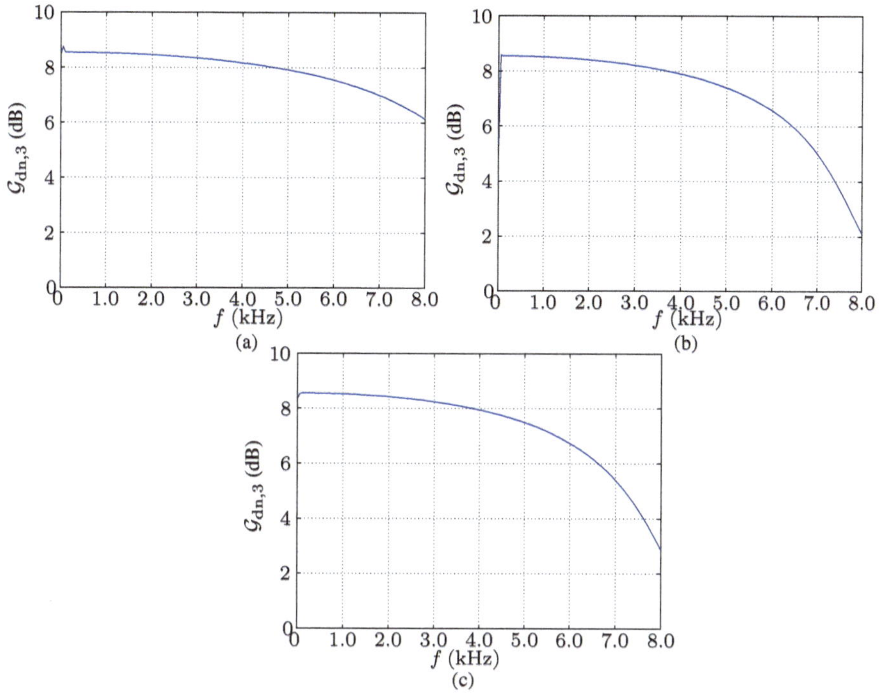


**Fig. 7.20** The directivity factor of a third-order CDMA with three distinct, as a function of frequency, for different values of $M$ and $\delta$: (a) $M = 7$, $\delta = 1$ cm, (b) $M = 9$, $\delta = 1$ cm, and (c) $M = 11$, $\delta = 8$ mm.

From all the simulations we have performed, we can definitely say that we can design good and robust first- and second-order CDMAs with the minimum-norm filter. However, the minimum-norm filter does not seem to be effective for the design of third-order CDMAs; more investigation is required here. One obvious possibility is to use concentric circular arrays.

## References


1. J. Benesty, J. Chen, and Y. Huang, *Microphone Array Signal Processing*. Berlin, Germany: Springer-Verlag, 2008.
2. G. W. Elko, "Superdirectional microphone arrays," in *Acoustic Signal Processing for Telecommunication*, S. L. Gay and J. Benesty, Eds. Boston, MA: Kluwer Academic Publishers, 2000, Chapter 10, pp. 181–237.
3. J. Benesty and J. Chen, *Study and Design of Differential Microphone Arrays*. Berlin, Germany: Springer-Verlag, 2012.
4. A. Booker and C. Y. Ong, "Multiple constraint adaptive filtering," *Geophysics*, vol. 36, pp. 498–509, June 1971.

5. O. Frost, "An algorithm for linearly constrained adaptive array processing," *Proc. IEEE*, vol. 60, pp. 926–935, Jan. 1972.
6. H. Cox, R. M. Zeskind, and M. M. Owen, "Robust adaptive beamforming," *IEEE Trans. Acoust., Speech, Signal Process.*, vol. ASSP-35, pp. 1365–1376, Oct. 1987.
7. J. Benesty and J. Chen, *Optimal Noise Reduction Filters–A Theoretical Study.* Springer Briefs in Electrical & Computer Engineering, 2011.
8. J. Benesty, J. Chen, and E. Habets, *Speech Enhancement in the STFT Domain.* Springer Briefs in Electrical & Computer Engineering, 2011.

# Chapter 8
# Design of Circular Differential Arrays with the Jacobi-Anger Expansion

In this chapter, we show that the patterns of differential arrays can be obtained from the general definition of the beampattern by approximating the exponential function with the Jacobi-Anger expansion. In other words, a directivity pattern of order $N$ can be obtained from the Jacobi-Anger expansion of order $N$, as long as this approximation holds. We also explain how to design circular differential arrays based on this approach and their relationship to adaptive beamforming. Finally, we discuss the ideal beampattern and demonstrate that it can be designed like a CDMA.

## 8.1 Beampattern Definitions Revisited

In this section, we show how the general definition of the beampattern with a UCA is very much related to the particular definition of an $N$th-order directivity pattern. We assume that $\theta_{\mathrm{s}} = 0$; the generalization to $\theta_{\mathrm{s}} \neq 0$ is straightforward. An alternative way to write the frequency-independent directivity pattern of an $N$th-order DMA is

$$\begin{aligned} \mathcal{B}_{\mathrm{Ch},N}(\theta) &= \sum_{n=0}^{N} b_{N,n} T_n(\cos\theta) \\ &= \sum_{n=0}^{N} b_{N,n} \cos(n\theta), \end{aligned} \tag{8.1}$$

where

$$T_n(\cos\theta) = \cos(n\theta), \ n = 0, 1, \ldots, N \tag{8.2}$$

are Chebyshev polynomials of the first kind [1], which have the recurrence relation:



J. Benesty et al., *Design of Circular Differential Microphone Arrays*,
Springer Topics in Signal Processing 12, DOI 10.1007/978-3-319-14842-7_8

$$T_{n+1}(\cos\theta) = 2\cos\theta \times T_n(\cos\theta) - T_{n-1}(\cos\theta) \tag{8.3}$$

with

$$T_0(\cos\theta) = 1,$$
$$T_1(\cos\theta) = \cos\theta,$$

and the real coefficients $b_{N,n}$, $n = 0, 1, \ldots, N$ of $\mathcal{B}_{\text{Ch},N}(\theta)$ can be easily determined from the coefficients $a_{N,n}$, $n = 0, 1, \ldots, N$ of $\mathcal{B}_N(\theta)$. There are other interesting ways to describe the pattern of an $N$th-order DMA [2], but the one proposed above is more appropriate in our context.

The Jacobi-Anger expansion [1], which represents an expansion of plane waves into a series of cylindrical waves, is

$$e^{\jmath\varpi\cos\theta} = J_0(\varpi) + 2\sum_{n=1}^{\infty} \jmath^n J_n(\varpi) T_n(\cos\theta), \tag{8.4}$$

where $J_n(\varpi)$ is the $n$th-order Bessel function of the first kind [1]. Assuming that $\varpi$ (or $r$) is small, we can approximate (8.4) up to its $N$th order, i.e.,

$$\begin{aligned} e^{\jmath\varpi\cos\theta} &\approx J_0(\varpi) + 2\sum_{n=1}^{N} \jmath^n J_n(\varpi)\cos(n\theta) \\ &= \sum_{n=0}^{N} J'_n(\varpi)\cos(n\theta), \end{aligned} \tag{8.5}$$

where

$$J'_n(\varpi) = \begin{cases} J_0(\varpi), & n = 0 \\ 2\jmath^n J_n(\varpi), & n = 1, 2, \ldots, N \end{cases}. \tag{8.6}$$

We recall that the beampattern of a UCA is

$$\mathcal{B}[\mathbf{h}(\omega), \theta] = \sum_{m=1}^{M} H_m^*(\omega) e^{\jmath\varpi\cos(\theta - \psi_m)}. \tag{8.7}$$

Using the symmetry of the filter's coefficients (see Chapter 2), we can rewrite (8.7) as

$$\begin{aligned} \mathcal{B}[\mathbf{h}(\omega), \theta] &= H_1^*(\omega) e^{\jmath\varpi\cos\theta} + H_{\mathcal{M}}^*(\omega) e^{-\jmath\varpi\cos\theta} + \\ &\sum_{m=2}^{\mathcal{M}-1} H_m^*(\omega)\left[e^{\jmath\varpi\cos(\theta - \psi_m)} + e^{\jmath\varpi\cos(\theta + \psi_m)}\right] \end{aligned} \tag{8.8}$$

for $M$ even or as

$$\mathcal{B}\left[\mathbf{h}(\omega), \theta\right] = \\ H_1^*(\omega) e^{\jmath\varpi\cos\theta} + \sum_{m=2}^{\mathcal{M}} H_m^*(\omega) \left[e^{\jmath\varpi\cos(\theta-\psi_m)} + e^{\jmath\varpi\cos(\theta+\psi_m)}\right] \tag{8.9}$$

for $M$ odd, where $\mathcal{M} = \lfloor \frac{M}{2} \rfloor + 1$. It is more convenient to express this beampattern in a compact form:

$$\begin{aligned}\mathcal{B}\left[\mathbf{h}(\omega), \theta\right] &= \sum_{m=1}^{\mathcal{M}} H_m^{\prime *}(\omega) \left[e^{\jmath\varpi\cos(\theta-\psi_m)} + e^{\jmath\varpi\cos(\theta+\psi_m)}\right] \\ &= \mathcal{B}\left[\mathbf{h}'(\omega), \theta\right],\end{aligned} \tag{8.10}$$

where

$$\mathbf{h}'(\omega) = \left[\, H_1'(\omega)\; H_2'(\omega)\; \cdots \; H_{\mathcal{M}}'(\omega) \,\right]^T \tag{8.11}$$

is a vector of length $\mathcal{M}$, with

$$H_1'(\omega) = \frac{1}{2} H_1(\omega), \tag{8.12}$$

$$H_m'(\omega) = H_m(\omega), \; m = 2, 3, \ldots, \mathcal{M} - 1, \tag{8.13}$$

and

$$H_{\mathcal{M}}'(\omega) = \begin{cases} H_{\mathcal{M}}(\omega), & \text{if } M \text{ odd} \\ \frac{1}{2} H_{\mathcal{M}}(\omega), & \text{if } M \text{ even} \end{cases}. \tag{8.14}$$

Substituting (8.5) into (8.10) and using the trigonometric relation:

$$\cos(a \pm b) = \cos a \cos b \mp \sin a \sin b, \tag{8.15}$$

it is not hard to verify that

$$\begin{aligned}\mathcal{B}\left[\mathbf{h}(\omega), \theta\right] &= \sum_{n=0}^{N} \cos(n\theta) \left[2J_n'(\varpi) \sum_{m=1}^{\mathcal{M}} \cos(n\psi_m) H_m^{\prime *}(\omega)\right] \\ &= \sum_{n=0}^{N} b_{N,n} \cos(n\theta),\end{aligned} \tag{8.16}$$

where

$$b_{N,n} = 2J_n'(\varpi) \sum_{m=1}^{\mathcal{M}} \cos(n\psi_m) H_m^{\prime *}(\omega), \; n = 0, 1, \ldots, N. \tag{8.17}$$

We observe from (8.16) that as long as $e^{\jmath\varpi\cos\theta}$ can be approximated by a Jacobi-Anger expansion of order $N$ (that is why the microphone spacing should be small), we can build $N$th-order differential arrays with circular apertures. We also observe from (8.17) that the gains $H_m(\omega)$, $m = 1, 2, \ldots, \mathcal{M}$, can be determined given the coefficients $b_{N,n}$, $n = 0, 1, \ldots, N$.

## 8.2 Design Principle

In the previous section, we demonstrated that there is an interesting relationship between the coefficients of a directivity pattern and the coefficients of a beamforming filter. Therefore, we can exploit this relationship to design CDMAs. In order to have a unique solution for (8.17), we must take $\mathcal{M} = N+1$; hence, $\lfloor \frac{M}{2} \rfloor = N$. As a result, it is easy to deduce the linear system of $N+1$ equations:

$$\boldsymbol{\Psi}_{N+1}\mathbf{h}'^{*}(\omega) = \mathbf{b}_{N+1}(\omega) \tag{8.18}$$

or

$$\boldsymbol{\Psi}_{N+1}\mathbf{h}'(\omega) = \mathbf{b}^{*}_{N+1}(\omega), \tag{8.19}$$

where

$$\boldsymbol{\Psi}_{N+1} = \begin{bmatrix} 1 & 1 & 1 & \cdots & 1 \\ 1 & \cos\psi_2 & \cos\psi_3 & \cdots & \cos\psi_{N+1} \\ 1 & \cos 2\psi_2 & \cos 2\psi_3 & \cdots & \cos 2\psi_{N+1} \\ \vdots & \vdots & \vdots & \ddots & \vdots \\ 1 & \cos N\psi_2 & \cos N\psi_3 & \cdots & \cos N\psi_{N+1} \end{bmatrix} \tag{8.20}$$

is a matrix of size $(N+1) \times (N+1)$,

$$\mathbf{h}'(\omega) = \begin{bmatrix} H'_1(\omega) \; H'_2(\omega) \; \cdots \; H'_{N+1}(\omega) \end{bmatrix}^T \tag{8.21}$$

is the filter of length $N+1$, and

$$\mathbf{b}^{*}_{N+1}(\omega) = \frac{1}{2}\begin{bmatrix} \frac{b_{N,0}}{J'^{*}_0(\varpi)} \; \frac{b_{N,1}}{J'^{*}_1(\varpi)} \; \cdots \; \frac{b_{N,N}}{J'^{*}_N(\varpi)} \end{bmatrix}^T \tag{8.22}$$

is a vector of length $N+1$ containing the theoretical coefficients of an $N$th-order DMA and the Bessel functions. Once we deduce $\mathbf{h}'(\omega)$ from (8.19), it is straightforward to derive the $N$th-order CDMA filter, $\mathbf{h}(\omega)$, of length $M$.

## 8.3 First-Order Differential Arrays

We need first to determine the coefficients $b_{1,n},\ n = 0, 1$ from $a_{1,n},\ n = 0, 1$. Since

$$\mathcal{B}_{\mathrm{Ch},1}(\theta) = b_{1,0} + b_{1,1}\cos\theta \tag{8.23}$$

and

$$\begin{aligned}\mathcal{B}_1(\theta) &= a_{1,0} + a_{1,1}\cos\theta \\ &= (1 - a_{1,1}) + a_{1,1}\cos\theta,\end{aligned} \tag{8.24}$$

it is obvious that

$$b_{1,0} = 1 - a_{1,1},\ b_{1,1} = a_{1,1}. \tag{8.25}$$

Therefore, (8.18) becomes

$$\begin{bmatrix} 1 & 1 \\ 1 & -\dfrac{1}{2} \end{bmatrix} \begin{bmatrix} H_1'(\omega) \\ H_2'(\omega) \end{bmatrix} = \begin{bmatrix} \dfrac{1 - a_{1,1}}{J_0(\varpi)} \\ \dfrac{\jmath a_{1,1}}{2J_1(\varpi)} \end{bmatrix}. \tag{8.26}$$

We deduce that

$$H_1(\omega) = \frac{1}{3}\left[\frac{1 - a_{1,1}}{J_0(\varpi)} + \jmath\frac{a_{1,1}}{J_1(\varpi)}\right] \tag{8.27}$$

and

$$H_2(\omega) = \frac{1}{3}\left[\frac{1 - a_{1,1}}{J_0(\varpi)} - \jmath\frac{a_{1,1}}{2J_1(\varpi)}\right]. \tag{8.28}$$

Finally, the first-order CDMA filter is

$$\mathbf{h}(\omega) = \left[ H_1(\omega)\ H_2(\omega)\ H_2(\omega) \right]^T, \tag{8.29}$$

which requires three microphones positioned on the vertices of an equilateral triangle. We can then plot the beampattern and compute the directivity factor as well as the white noise gain.

Let us take the example of the first-order cardioid, i.e., $a_{1,1} = 1/2$.

Figure 8.1 shows the patterns of the first-order cardioid for several frequencies and two values of $\delta$ (1 and 2 cm).

Figure 8.2 gives plots of the directivity factor of the first-order cardioid, as a function of frequency, for different values of $\delta$.

Figure 8.3 displays plots of the white noise gain of the first-order cardioid, as a function of frequency, for different values of $\delta$.

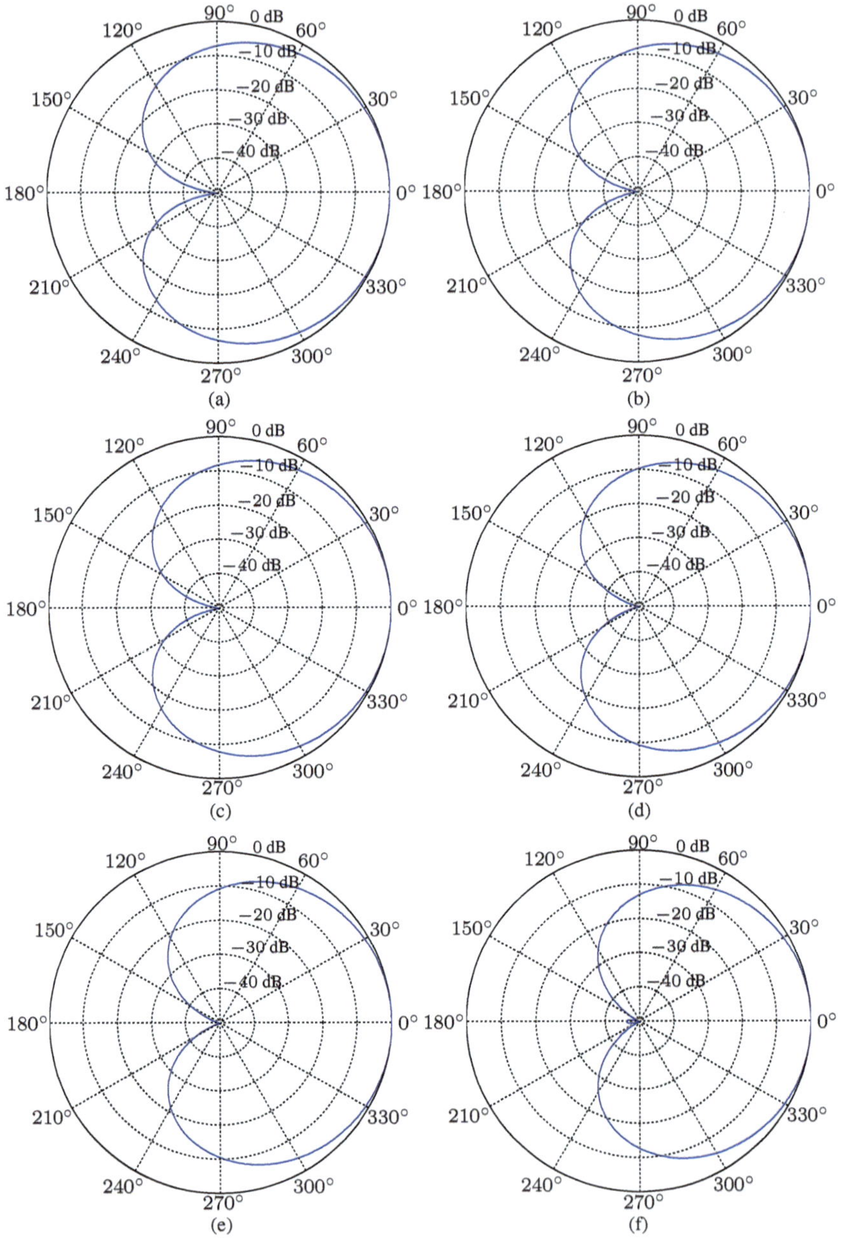


**Fig. 8.1** Patterns of the first-order cardioid (with three microphones) for several frequencies and two values of $\delta$: (a) $f = 0.5$ kHz, $\delta = 1$ cm, (b) $f = 0.5$ kHz, $\delta = 2$ cm, (c) $f = 1$ kHz, $\delta = 1$ cm, (d) $f = 1$ kHz, $\delta = 2$ cm, (e) $f = 3$ kHz, $\delta = 1$ cm, and (f) $f = 3$ kHz, $\delta = 2$ cm.

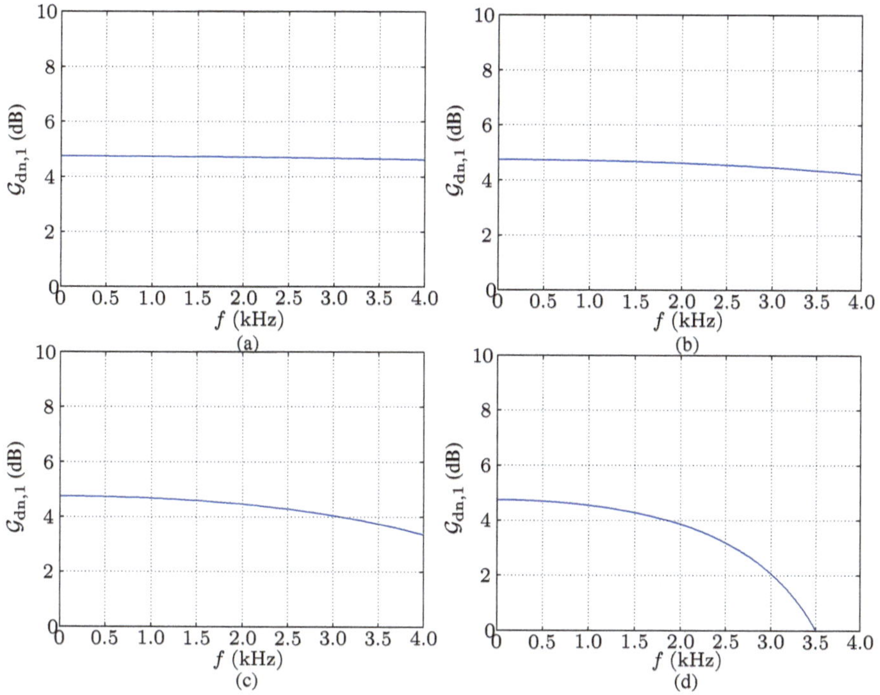


**Fig. 8.2** The directivity factor of the first-order cardioid (with three microphones), as a function of frequency, for different values of $\delta$: (a) $\delta = 1$ cm, (b) $\delta = 2$ cm, (c) $\delta = 3$ cm, and (d) $\delta = 5$ cm.

The performance of this approach is similar to the one proposed in Chapter 3. It is of interest to observe, though, that here the directivity factor is better in high frequencies.

## 8.4 Second-Order Differential Arrays

Since

$$\begin{aligned}\mathcal{B}_2(\theta) &= a_{2,0} + a_{2,1}\cos\theta + a_{2,2}\cos^2\theta \\ &= (1 - a_{2,1} - a_{2,2}) + a_{2,1}\cos\theta + a_{2,2}\cos^2\theta \end{aligned} \tag{8.30}$$

and

$$\begin{aligned}\mathcal{B}_{\rm Ch,2}(\theta) &= b_{2,0} + b_{2,1}\cos\theta + b_{2,2}\cos(2\theta) \\ &= (b_{2,0} - b_{2,2}) + b_{2,1}\cos\theta + 2b_{2,2}\cos^2\theta, \end{aligned} \tag{8.31}$$

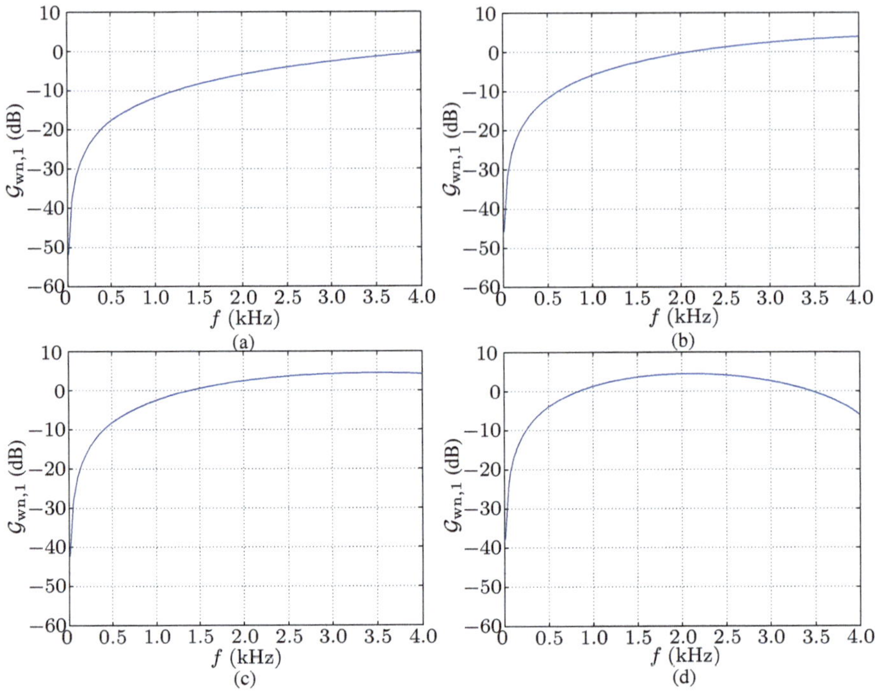


**Fig. 8.3** The white noise gain of the first-order cardioid (with three microphones), as a function of frequency, for different values of $\delta$: (a) $\delta = 1$ cm, (b) $\delta = 2$ cm, (c) $\delta = 3$ cm, and (d) $\delta = 5$ cm.

we deduce that

$$b_{2,0} = 1 - a_{2,1} - \frac{a_{2,2}}{2}, \; b_{2,1} = a_{2,1}, \; b_{2,2} = \frac{a_{2,2}}{2}. \tag{8.32}$$

We have two possibilities to design a second-order CDMA: $M = 4$ and $M = 5$.

For $M = 4$, we need to solve the linear system of equations:

$$\begin{bmatrix} 1 & 1 & 1 \\ 1 & 0 & -1 \\ 1 & -1 & 1 \end{bmatrix} \begin{bmatrix} H_1'(\omega) \\ H_2'(\omega) \\ H_3'(\omega) \end{bmatrix} = \frac{1}{2} \begin{bmatrix} \frac{b_{2,0}}{J_0'^*(\varpi)} \\ \frac{b_{2,1}}{J_1'^*(\varpi)} \\ \frac{b_{2,2}}{J_2'^*(\varpi)} \end{bmatrix}. \tag{8.33}$$

Then, it is straightforward to find the second-order CDMA filter:

$$H_1(\omega) = 2H_1'(\omega), \; H_2(\omega) = H_2'(\omega), \; H_3(\omega) = 2H_3'(\omega), \tag{8.34}$$

$$\mathbf{h}(\omega) = \left[ H_1(\omega) \; H_2(\omega) \; H_3(\omega) \; H_2(\omega) \right]^T, \tag{8.35}$$

which requires four microphones positioned on the vertices of a square.

To illustrate how this CDMA works, we take the example of the second-order cardioid, i.e., $a_{2,2} = a_{2,1} = 1/2$.

Figure 8.4 shows the patterns of the second-order cardioid (with four microphones) for several frequencies and two values of $\delta$ (1 and 2 cm).

Figure 8.5 gives plots of the directivity factor of the second-order cardioid (with four microphones), as a function of frequency, for different values of $\delta$.

Figure 8.6 displays plots of the white noise gain of the second-order cardioid (with four microphones), as a function of frequency, for different values of $\delta$.

For $M = 5$, the linear system of equations to solve is

$$\begin{bmatrix} 1 & 1 & 1 \\ 1 & 0.31 & -0.81 \\ 1 & -0.81 & 0.31 \end{bmatrix} \begin{bmatrix} H_1'(\omega) \\ H_2'(\omega) \\ H_3'(\omega) \end{bmatrix} = \frac{1}{2} \begin{bmatrix} \frac{b_{2,0}}{J_0'^*(\varpi)} \\ \frac{b_{2,1}}{J_1'^*(\varpi)} \\ \frac{b_{2,2}}{J_2'^*(\varpi)} \end{bmatrix}. \tag{8.36}$$

Therefore, the second-order CDMA filter is

$$H_1(\omega) = 2H_1'(\omega), \; H_2(\omega) = H_2'(\omega), \; H_3(\omega) = H_3'(\omega), \tag{8.37}$$

$$\mathbf{h}(\omega) = \begin{bmatrix} H_1(\omega) \; H_2(\omega) \; H_3(\omega) \; H_3(\omega) \; H_2(\omega) \end{bmatrix}^T, \tag{8.38}$$

which requires five microphones positioned on the vertices of a regular pentagon.

Figure 8.7 shows the patterns of the second-order cardioid (with five microphones) for several frequencies and two values of $\delta$ (1 and 2 cm).

Figure 8.8 gives plots of the directivity factor of the second-order cardioid (with five microphones), as a function of frequency, for different values of $\delta$.

Figure 8.9 displays plots of the white noise gain of the second-order cardioid (with five microphones), as a function of frequency, for different values of $\delta$.

The performance of the approach presented here (with either 4 or 5 microphones) is comparable to the one developed in Chapter 4.

## 8.5 Third-Order Differential Arrays

Since

$$\begin{aligned} \mathcal{B}_3(\theta) &= a_{3,0} + a_{3,1}\cos\theta + a_{3,2}\cos^2\theta + a_{3,3}\cos^3\theta \qquad (8.39) \\ &= (1 - a_{3,1} - a_{3,2} - a_{3,3}) + a_{3,1}\cos\theta + a_{3,2}\cos^2\theta + a_{3,3}\cos^3\theta \end{aligned}$$

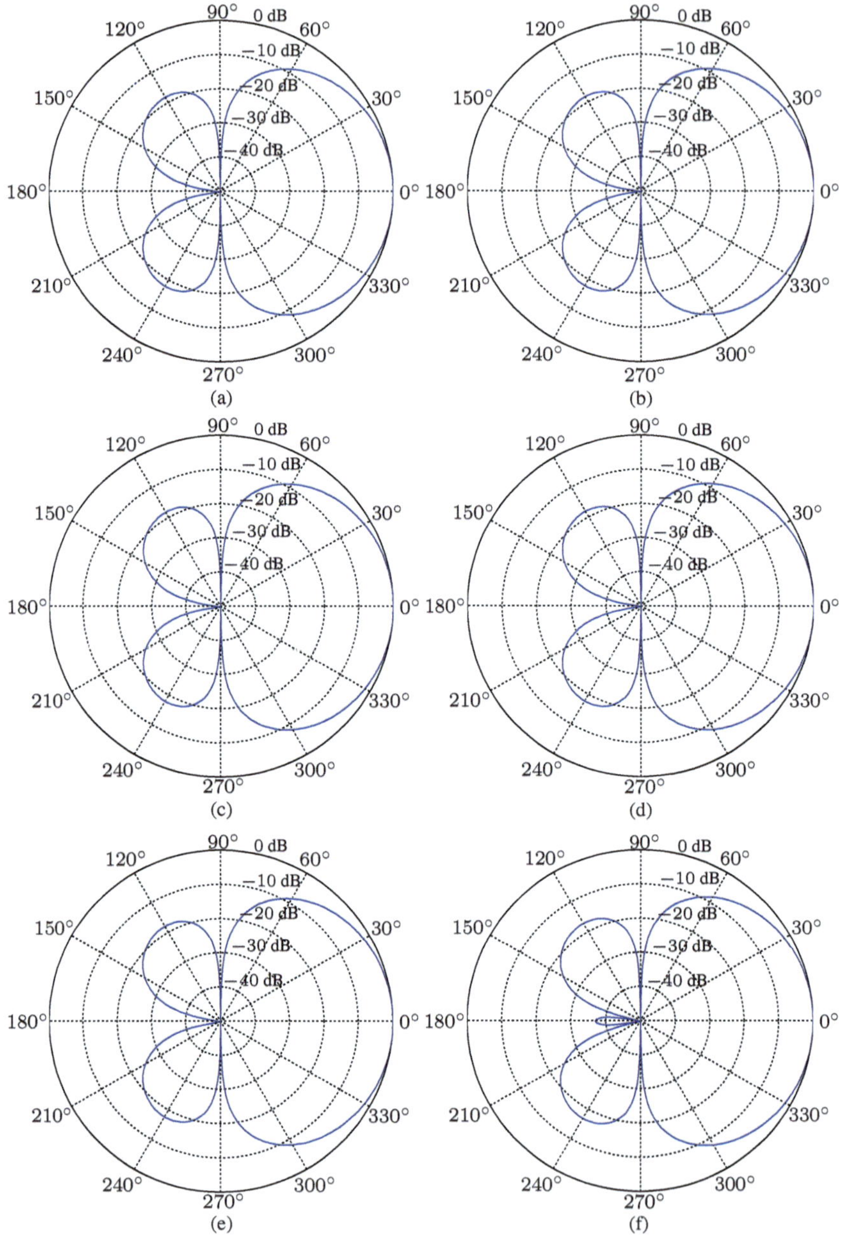


**Fig. 8.4** Patterns of the second-order cardioid (with four microphones) for several frequencies and two values of $\delta$: (a) $f = 0.5$ kHz, $\delta = 1$ cm, (b) $f = 0.5$ kHz, $\delta = 2$ cm, (c) $f = 1$ kHz, $\delta = 1$ cm, (d) $f = 1$ kHz, $\delta = 2$ cm, (e) $f = 3$ kHz, $\delta = 1$ cm, and (f) $f = 3$ kHz, $\delta = 2$ cm.

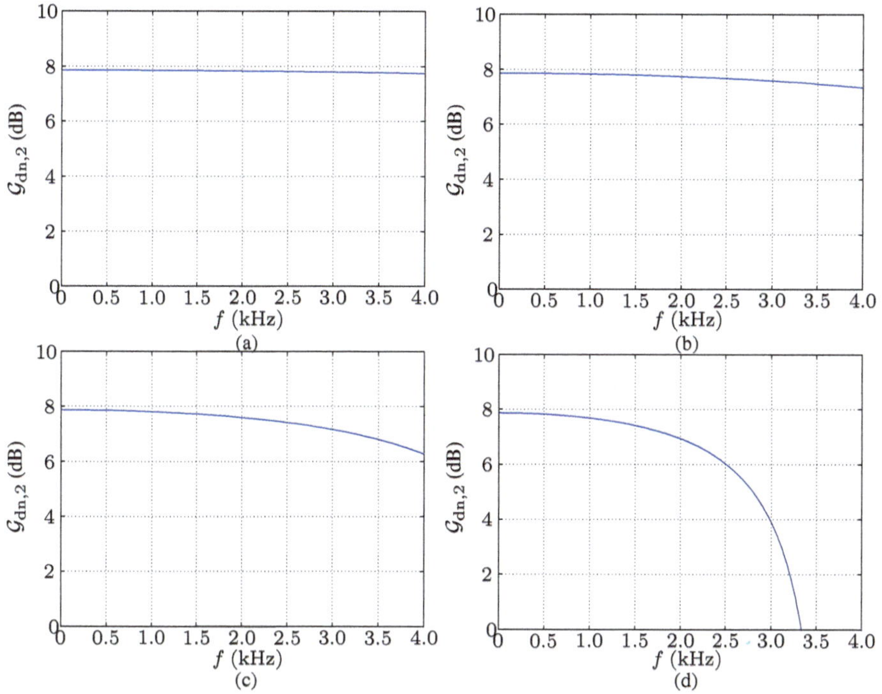


**Fig. 8.5** The directivity factor of the second-order cardioid (with four microphones), as a function of frequency, for different values of $\delta$: (a) $\delta = 1$ cm, (b) $\delta = 2$ cm, (c) $\delta = 3$ cm, and (d) $\delta = 5$ cm.

and

$$\begin{aligned}\mathcal{B}_{\mathrm{Ch},3}(\theta) &= b_{3,0} + b_{3,1}\cos\theta + b_{3,2}\cos(2\theta) + b_{3,3}\cos(3\theta) \\ &= (b_{3,0} - b_{3,2}) + (b_{3,1} - 3b_{3,3})\cos\theta + 2b_{3,2}\cos^2\theta + 4b_{3,3}\cos^3\theta,\end{aligned} \tag{8.40}$$

we find that

$$b_{3,0} = 1 - a_{3,1} - \frac{a_{3,2}}{2} - a_{3,3},\ b_{3,1} = \frac{3}{4}a_{3,3} + a_{3,1},\ b_{3,2} = \frac{a_{3,2}}{2},\ b_{3,3} = \frac{a_{3,3}}{4}. \tag{8.41}$$

To design a third-order CDMA, we have two possibilities: $M = 6$ and $M = 7$.

For $M = 6$, we need to solve the linear system of equations:

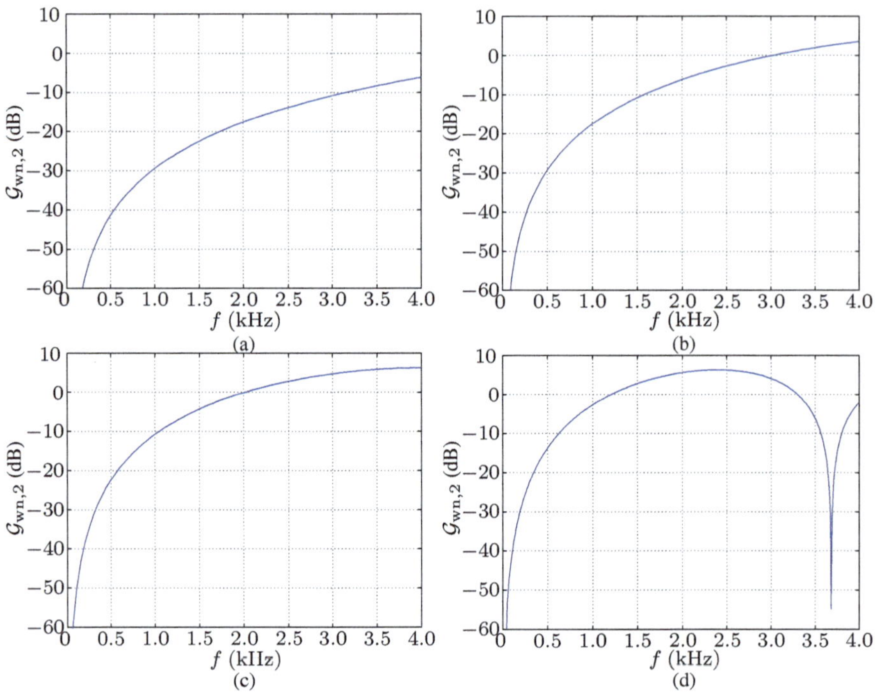


**Fig. 8.6** The white noise gain of the second-order cardioid (with four microphones), as a function of frequency, for different values of $\delta$: (a) $\delta = 1$ cm, (b) $\delta = 2$ cm, (c) $\delta = 3$ cm, and (d) $\delta = 5$ cm.

$$\begin{bmatrix} 1 & 1 & 1 & 1 \\ 1 & 0.5 & -0.5 & -1 \\ 1 & -0.5 & -0.5 & 1 \\ 1 & -1 & 1 & -1 \end{bmatrix} \begin{bmatrix} H_1'(\omega) \\ H_2'(\omega) \\ H_3'(\omega) \\ H_4'(\omega) \end{bmatrix} = \frac{1}{2} \begin{bmatrix} \frac{b_{3,0}}{J_0'^{*}(\varpi)} \\ \frac{b_{3,1}}{J_1'^{*}(\varpi)} \\ \frac{b_{3,2}}{J_2'^{*}(\varpi)} \\ \frac{b_{3,3}}{J_3'^{*}(\varpi)} \end{bmatrix}. \tag{8.42}$$

Once the previous system solved, we easily deduce the third-order CDMA filter:

$$H_1(\omega) = 2H_1'(\omega), \; H_2(\omega) = H_2'(\omega), \; H_3(\omega) = H_3'(\omega), \; H_4(\omega) = 2H_4'(\omega), \tag{8.43}$$

$$\mathbf{h}(\omega) = \left[ H_1(\omega) \; H_2(\omega) \; H_3(\omega) \; H_4(\omega) \; H_3(\omega) \; H_2(\omega) \right]^T, \tag{8.44}$$

which requires six microphones positioned on the vertices of a regular hexagon.

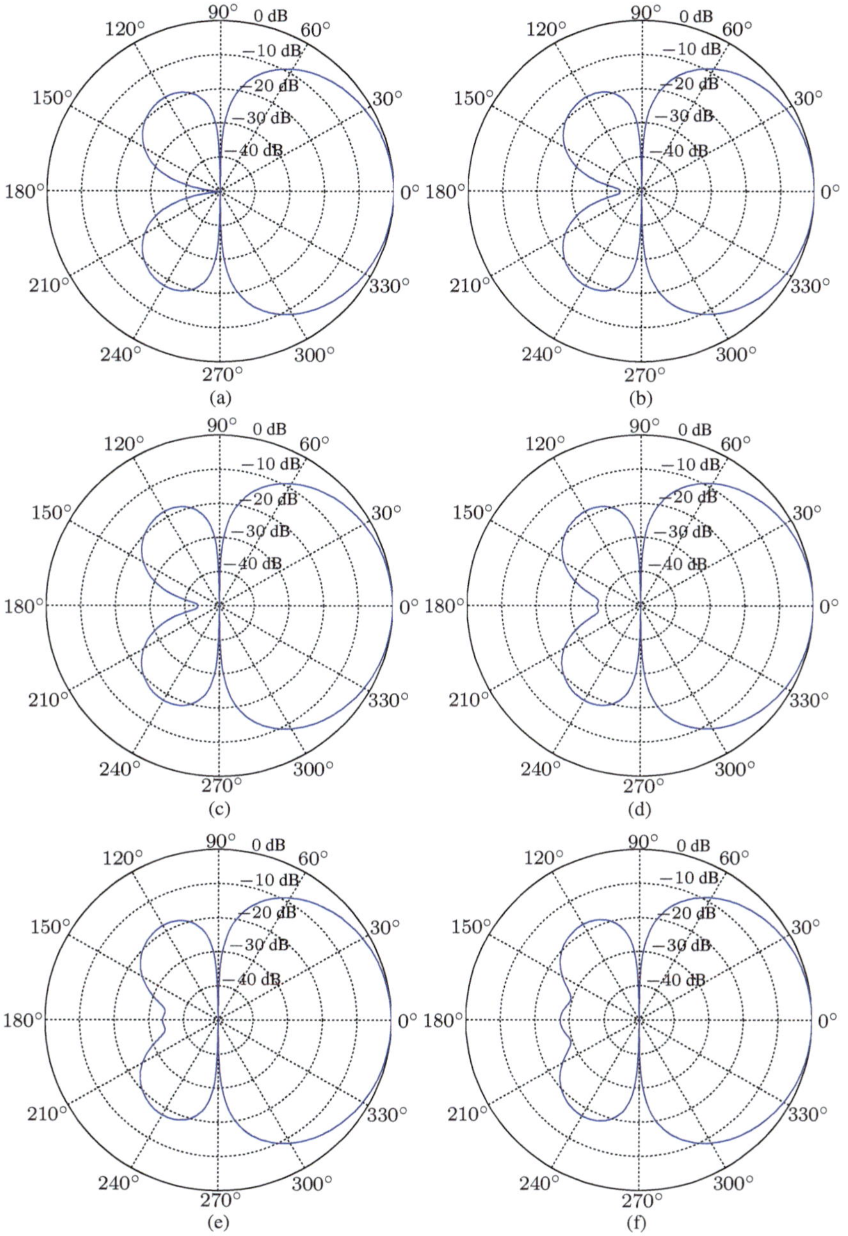


**Fig. 8.7** Patterns of the second-order cardioid (with five microphones) for several frequencies and two values of $\delta$: (a) $f = 0.5$ kHz, $\delta = 1$ cm, (b) $f = 0.5$ kHz, $\delta = 2$ cm, (c) $f = 1$ kHz, $\delta = 1$ cm, (d) $f = 1$ kHz, $\delta = 2$ cm, (e) $f = 3$ kHz, $\delta = 1$ cm, and (f) $f = 3$ kHz, $\delta = 2$ cm.

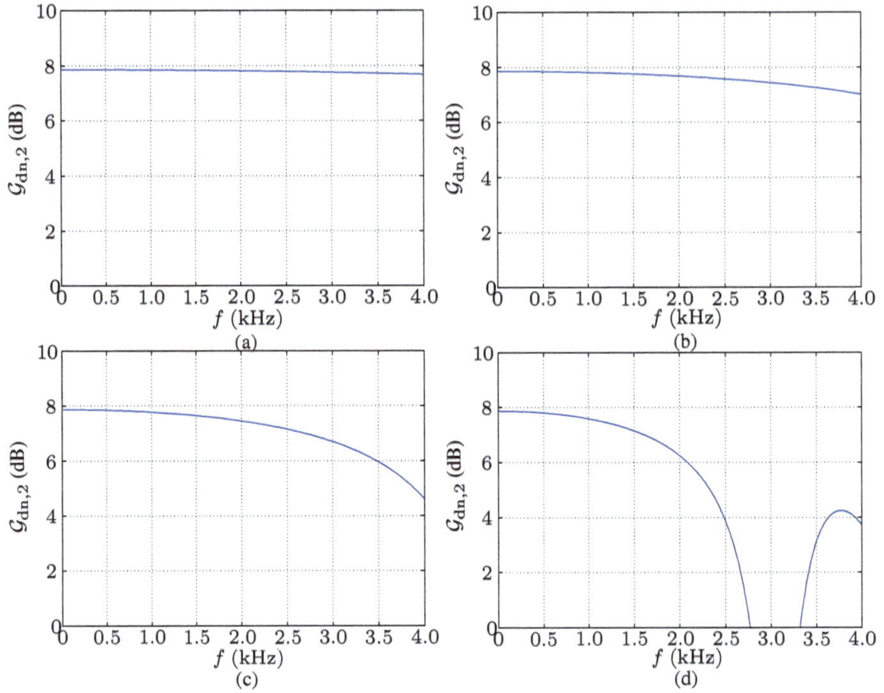


**Fig. 8.8** The directivity factor of the second-order cardioid (with five microphones), as a function of frequency, for different values of $\delta$: (a) $\delta = 1$ cm, (b) $\delta = 2$ cm, (c) $\delta = 3$ cm, and (d) $\delta = 5$ cm.

Let us take the example of the third-order cardioid, i.e., $a_{3,3} = a_{3,2} = 1/2$ and $a_{3,1} = 0$.

Figure 8.10 shows the patterns of the third-order cardioid (with six microphones) for several frequencies and two values of $\delta$ (1 and 2 cm).

Figure 8.11 gives plots of the directivity factor of the third-order cardioid (with six microphones), as a function of frequency, for different values of $\delta$.

Figure 8.12 displays plots of the white noise gain of the third-order cardioid (with six microphones), as a function of frequency, for different values of $\delta$.

For $M = 7$, the linear system of equations to be solved is

$$\begin{bmatrix} 1 & 1 & 1 & 1 \\ 1 & 0.62 & -0.22 & -0.9 \\ 1 & -0.22 & -0.9 & 0.62 \\ 1 & -0.9 & 0.62 & -0.22 \end{bmatrix} \begin{bmatrix} H_1'(\omega) \\ H_2'(\omega) \\ H_3'(\omega) \\ H_4'(\omega) \end{bmatrix} = \frac{1}{2} \begin{bmatrix} \frac{b_{3,0}}{J_0'^*(\varpi)} \\ \frac{b_{3,1}}{J_1'^*(\varpi)} \\ \frac{b_{3,2}}{J_2'^*(\varpi)} \\ \frac{b_{3,3}}{J_3'^*(\varpi)} \end{bmatrix}. \tag{8.45}$$

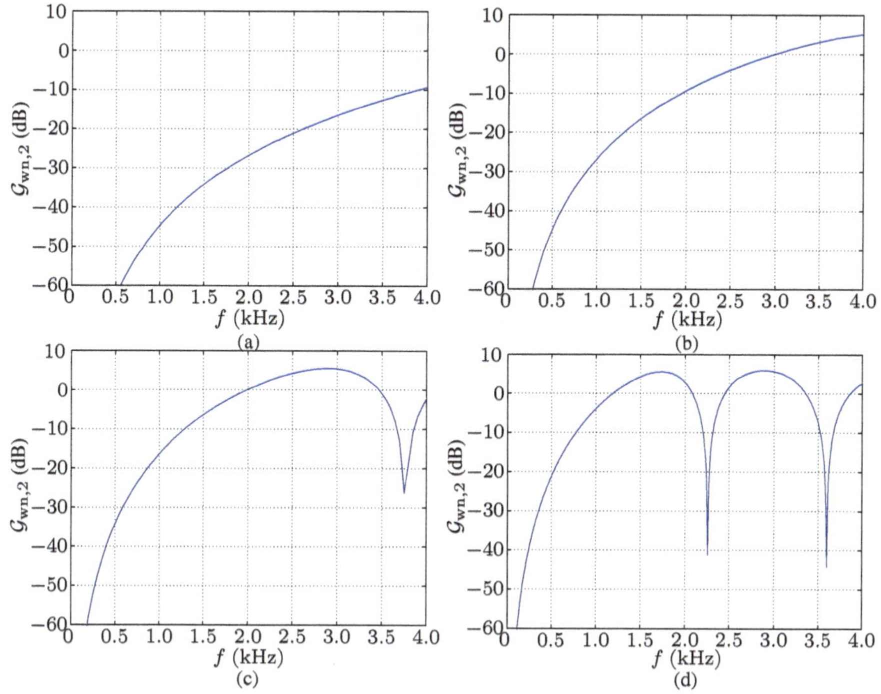


**Fig. 8.9** The white noise gain of the second-order cardioid (with five microphones), as a function of frequency, for different values of $\delta$: (a) $\delta = 1$ cm, (b) $\delta = 2$ cm, (c) $\delta = 3$ cm, and (d) $\delta = 5$ cm.

Therefore, the third-order CDMA filter is

$$H_1(\omega) = 2H_1'(\omega), \; H_2(\omega) = H_2'(\omega), \; H_3(\omega) = H_3'(\omega), \; H_4(\omega) = H_4'(\omega), \quad (8.46)$$

$$\mathbf{h}(\omega) = \left[ H_1(\omega) \; H_2(\omega) \; H_3(\omega) \; H_4(\omega) \; H_4(\omega) \; H_3(\omega) \; H_2(\omega) \right]^T, \quad (8.47)$$

which requires seven microphones positioned on the vertices of a regular heptagon.

Figure 8.13 shows the patterns of the third-order cardioid (with seven microphones) for several frequencies and two values of $\delta$ (1 and 2 cm).

Figure 8.14 gives plots of the directivity factor of the third-order cardioid (with seven microphones), as a function of frequency, for different values of $\delta$.

Figure 8.15 displays plots of the white noise gain of the third-order cardioid (with seven microphones), as a function of frequency, for different values of $\delta$.

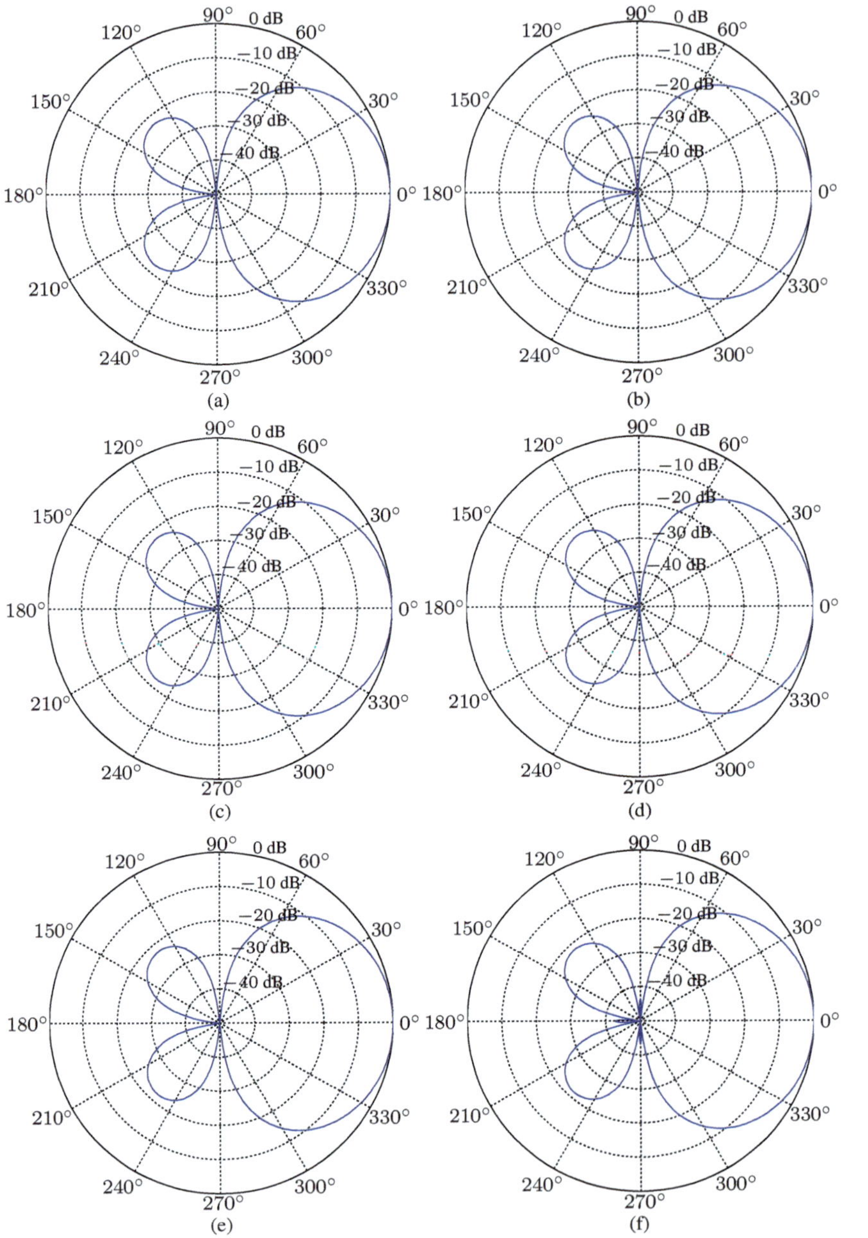


**Fig. 8.10** Patterns of the third-order cardioid (with six microphones) for several frequencies and two values of $\delta$: (a) $f = 0.5$ kHz, $\delta = 1$ cm, (b) $f = 0.5$ kHz, $\delta = 2$ cm, (c) $f = 1$ kHz, $\delta = 1$ cm, (d) $f = 1$ kHz, $\delta = 2$ cm, (e) $f = 3$ kHz, $\delta = 1$ cm, and (f) $f = 3$ kHz, $\delta = 2$ cm.

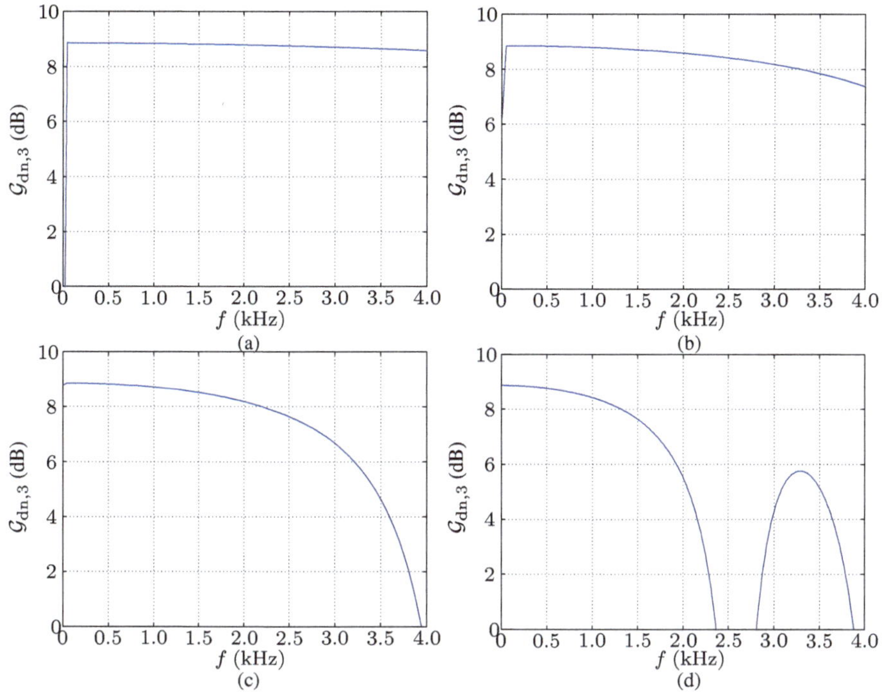


**Fig. 8.11** The directivity factor of the third-order cardioid (with six microphones), as a function of frequency, for different values of $\delta$: (a) $\delta = 1$ cm, (b) $\delta = 2$ cm, (c) $\delta = 3$ cm, and (d) $\delta = 5$ cm.

## 8.6 Adaptive Beamforming versus Differential Arrays

The linear system given in (8.18) can be generalized to

$$\mathbf{\Psi}_{N+1}\mathbf{h}'(\omega) = \mathbf{b}^*_{N+1}(\omega), \tag{8.48}$$

where now

$$\mathbf{\Psi}_{N+1} = \begin{bmatrix} 1 & 1 & 1 & \cdots & 1 \\ 1 & \cos\psi_2 & \cos\psi_3 & \cdots & \cos\psi_{\mathcal{M}} \\ 1 & \cos 2\psi_2 & \cos 2\psi_3 & \cdots & \cos 2\psi_{\mathcal{M}} \\ \vdots & \vdots & \vdots & \ddots & \vdots \\ 1 & \cos N\psi_2 & \cos N\psi_3 & \cdots & \cos N\psi_{\mathcal{M}} \end{bmatrix} \tag{8.49}$$

is a matrix of size $(N+1)\times\mathcal{M}$, with $\mathcal{M} = \lfloor \frac{M}{2} \rfloor + 1$,

$$\mathbf{h}'(\omega) = \begin{bmatrix} H'_1(\omega) \; H'_2(\omega) \; \ldots \; H'_{\mathcal{M}}(\omega) \end{bmatrix}^T \tag{8.50}$$

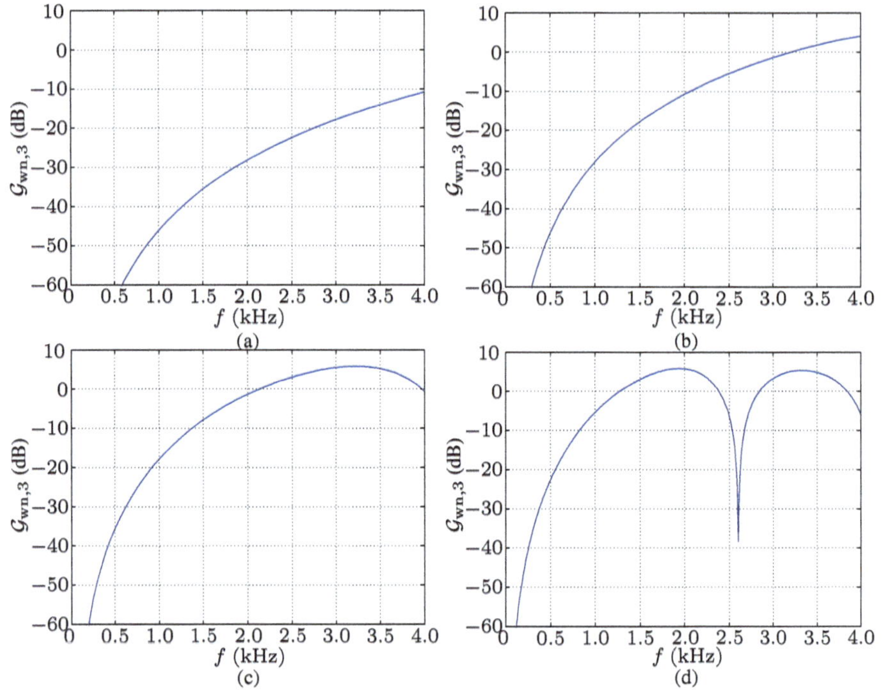


**Fig. 8.12** The white noise gain of the third-order cardioid (with six microphones), as a function of frequency, for different values of $\delta$: (a) $\delta = 1$ cm, (b) $\delta = 2$ cm, (c) $\delta = 3$ cm, and (d) $\delta = 5$ cm.

is the filter of length $\mathcal{M}$, and the vector $\mathbf{b}_{N+1}^{*}(\omega)$ was already defined in (8.22).

In adaptive beamforming [3], we minimize the residual noise at the beamformer output subject to the constraints summarized in (8.48). Mathematically, this is equivalent to

$$\min_{\mathbf{h}'(\omega)} \mathbf{h}'^{H}(\omega)\, \boldsymbol{\Phi}'_{\mathbf{v}}(\omega)\, \mathbf{h}'(\omega) \quad \text{subject to} \quad \boldsymbol{\Psi}_{N+1}\mathbf{h}'(\omega) = \mathbf{b}_{N+1}^{*}(\omega), \tag{8.51}$$

where $\boldsymbol{\Phi}'_{\mathbf{v}}(\omega)$ (of size $\mathcal{M} \times \mathcal{M}$) is built from $\boldsymbol{\Phi}_{\mathbf{v}}(\omega)$ (of size $M \times M$) to take into account the symmetry of $\mathbf{h}(\omega)$. We easily deduce that the solution is the LCMV filter [3], [4]:

$$\mathbf{h}'_{\mathrm{LCMV}}(\omega) = \boldsymbol{\Phi}'^{-1}_{\mathbf{v}}(\omega)\, \boldsymbol{\Psi}_{N+1}^{T} \left[ \boldsymbol{\Psi}_{N+1} \boldsymbol{\Phi}'^{-1}_{\mathbf{v}}(\omega)\, \boldsymbol{\Psi}_{N+1}^{T} \right]^{-1} \mathbf{b}_{N+1}^{*}(\omega). \tag{8.52}$$

We observe that for the matrix $\boldsymbol{\Psi}_{N+1} \boldsymbol{\Phi}'^{-1}_{\mathbf{v}}(\omega)\, \boldsymbol{\Psi}_{N+1}^{T}$ in (8.52) to be full rank, we must have $N + 1 \leq \mathcal{M} = \lfloor \frac{M}{2} \rfloor + 1$, which is the same condition to design a differential array of order $N$ with a UCA.

For $\mathcal{M} = N + 1$, we easily deduce from (8.52) that

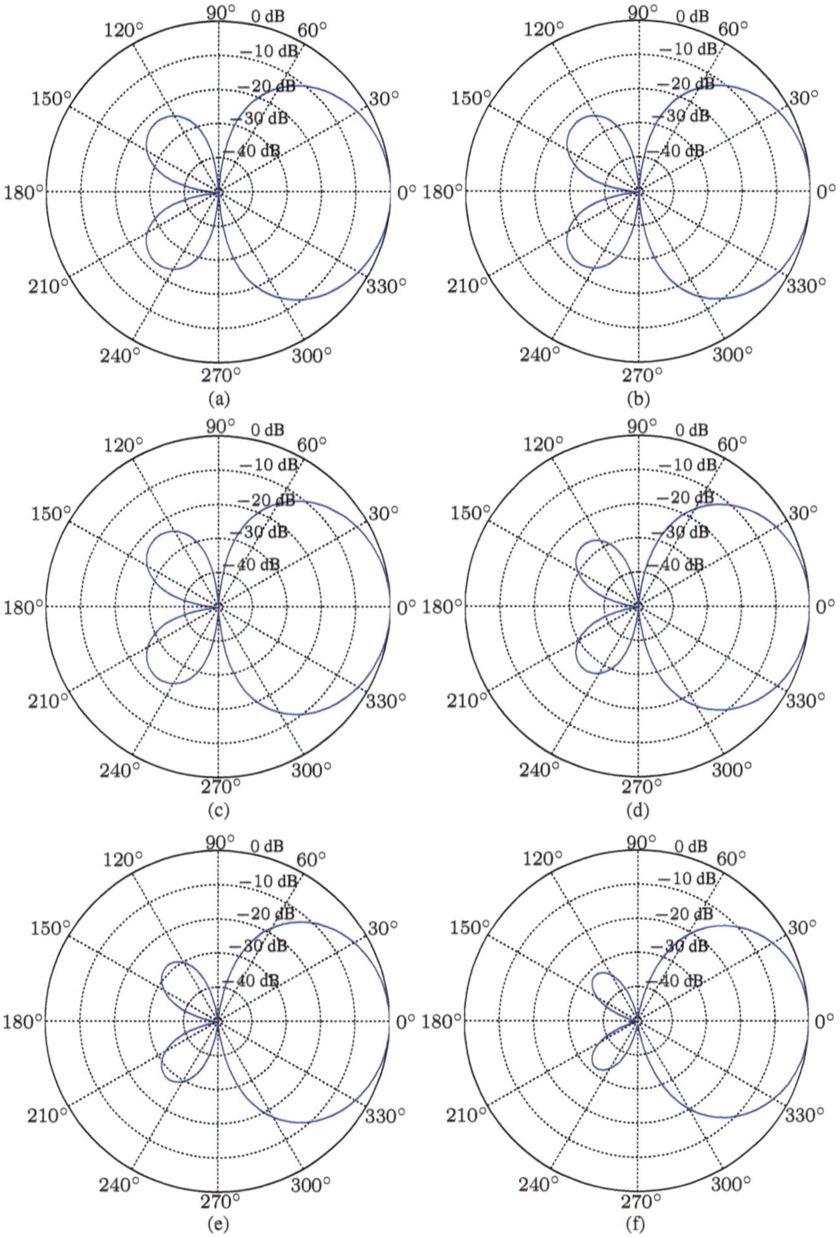


**Fig. 8.13** Patterns of the third-order cardioid (with seven microphones) for several frequencies and two values of $\delta$: (a) $f = 0.5$ kHz, $\delta = 1$ cm, (b) $f = 0.5$ kHz, $\delta = 2$ cm, (c) $f = 1$ kHz, $\delta = 1$ cm, (d) $f = 1$ kHz, $\delta = 2$ cm, (e) $f = 3$ kHz, $\delta = 1$ cm, and (f) $f = 3$ kHz, $\delta = 2$ cm.

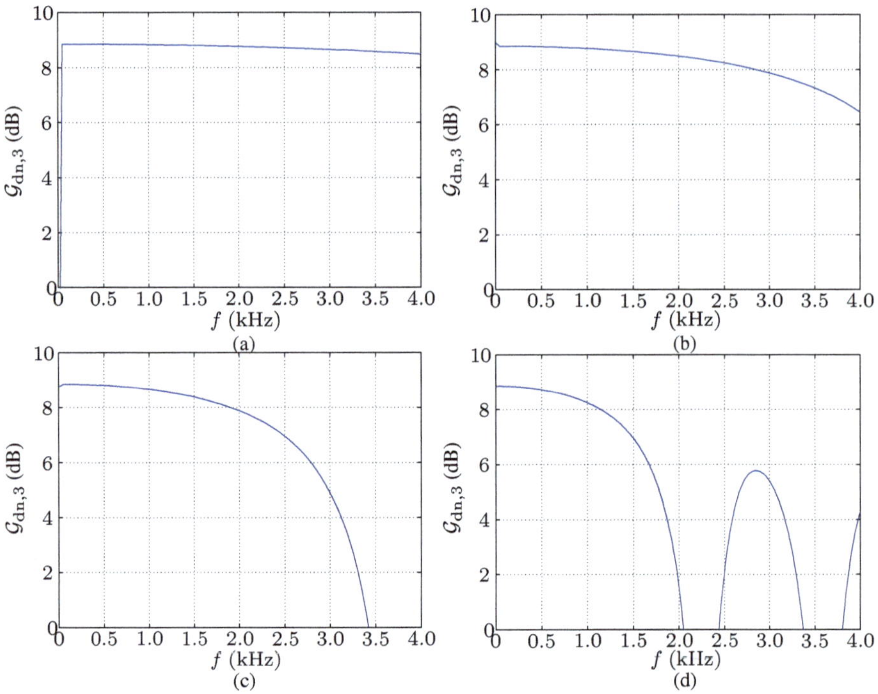


**Fig. 8.14** The directivity factor of the third-order cardioid (with seven microphones), as a function of frequency, for different values of $\delta$: (a) $\delta = 1$ cm, (b) $\delta = 2$ cm, (c) $\delta = 3$ cm, and (d) $\delta = 5$ cm.

$$\mathbf{h}'_{\text{LCMV}}(\omega) = \boldsymbol{\Psi}_{N+1}^{-1}\mathbf{b}_{N+1}^{*}(\omega), \tag{8.53}$$

which corresponds exactly to the filter of an $N$th-order CDMA or the solution of (8.48).

For $\mathcal{M} > N + 1$ and spatially white noise, (8.52) becomes

$$\mathbf{h}'_{\text{LCMV}}(\omega) = \boldsymbol{\Psi}_{N+1}^{T}\left(\boldsymbol{\Psi}_{N+1}\boldsymbol{\Psi}_{N+1}^{T}\right)^{-1}\mathbf{b}_{N+1}^{*}(\omega), \tag{8.54}$$

which corresponds to the minimum-norm solution of (8.48). This shows that the LCMV filter is fundamentally related to the filter of an $N$th-order CDMA. The minimum-norm filter given in (8.54) can also be a good solution against white noise amplification.

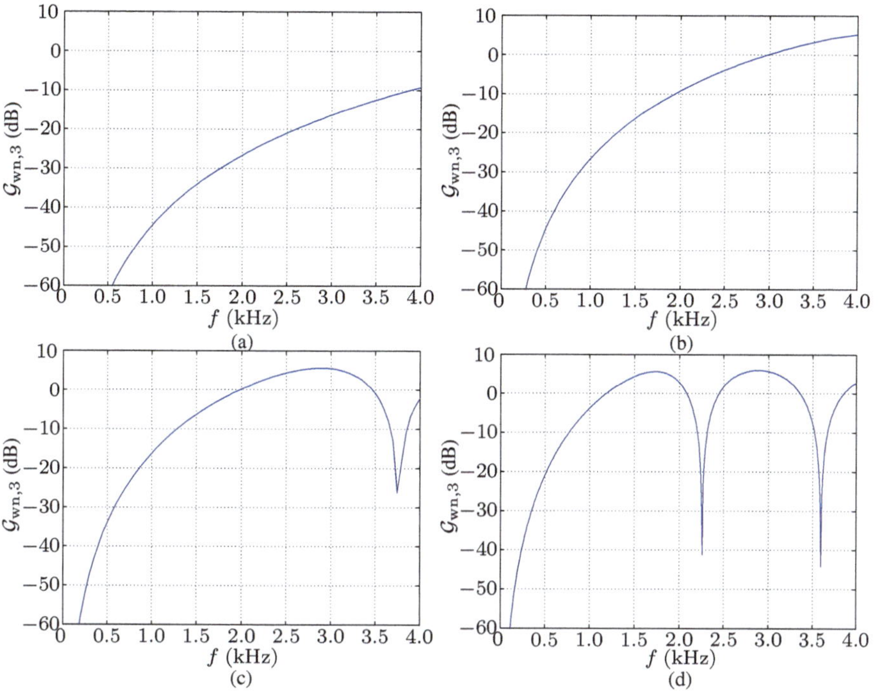


**Fig. 8.15** The white noise gain of the third-order cardioid (with seven microphones), as a function of frequency, for different values of $\delta$: (a) $\delta = 1$ cm, (b) $\delta = 2$ cm, (c) $\delta = 3$ cm, and (d) $\delta = 5$ cm.

## 8.7 Design of the Ideal Beampattern

The ideal beampattern has an infinitely high spike at $\theta = \theta_s$ and zero everywhere else [5]. Therefore, it coincides with the Dirac's delta function, i.e.,

$$\mathcal{B}_{Id}(\theta - \theta_s) = \delta_D(\theta - \theta_s) \tag{8.55}$$
$$= \begin{cases} +\infty, & \theta = \theta_s \\ 0, & \theta \neq \theta_s \end{cases}.$$

The Fourier series expansion of the Dirac's delta function is well known [6]. We deduce that

$$\mathcal{B}_{Id}(\theta - \theta_s) = \frac{1}{2\pi} + \frac{1}{\pi}\sum_{n=1}^{\infty} \cos[n(\theta - \theta_s)]. \tag{8.56}$$

If we approximate (8.56) up to its $N$th order, we get

$$\mathcal{B}_{\mathrm{Id},N}\left(\theta-\theta_{\mathrm{s}}\right)=\sum_{n=0}^{N} b_{N,n} \cos\left[n\left(\theta-\theta_{\mathrm{s}}\right)\right], \tag{8.57}$$

where

$$b_{N,n}=\begin{cases}\dfrac{1}{2\pi}, & n=0 \\ \dfrac{1}{\pi}, & n=1,2,\ldots,N\end{cases}. \tag{8.58}$$

For $\theta_{\mathrm{s}}=0$, we see that $\mathcal{B}_{\mathrm{Id},N}(\theta)$ is a particular case of $\mathcal{B}_{\mathrm{Ch},N}(\theta)$. As a consequence, designing an $N$th-order ideal beampattern with a UCA is equivalent to designing an $N$th-order CDMA.

## References


1. M. Abramowitz and I. A. Stegun, Eds., *Handbook of Mathematical Functions with Formulas, Graphs, and Mathematical Tables, 9th Printing.* New York: Dover, 1972.
2. T. D. Abhayapala and A. Gupta, "Higher order differential-integral microphone arrays," *J. Acoust. Soc. Am.*, vol. 127, pp. EL227–EL233, May 2010.
3. J. Benesty, J. Chen, and Y. Huang, *Microphone Array Signal Processing.* Berlin, Germany: Springer-Verlag, 2008.
4. O. Frost, "An algorithm for linearly constrained adaptive array processing," *Proc. IEEE*, vol. 60, pp. 926–935, Jan. 1972.
5. Z. Li, R. Duraiswami, and L. S. Davis, "Recording and reproducing high order surround auditory scenes for mixed and augmented reality," in *Proc. IEEE & ACM ISMAR*, 2004.
6. R. Bracewell, *The Fourier Transform and Its Applications, 3rd ed.* New York: McGraw-Hill, 2000.

# Index




J. Benesty et al., *Design of Circular Differential Microphone Arrays*,
Springer Topics in Signal Processing 12, DOI 10.1007/978-3-319-14842-7

MIX
Papier aus verantwortungsvollen Quellen
Paper from responsible sources
FSC® C105338

If you have any concerns about our products,
you can contact us on
**ProductSafety@springernature.com**

In case Publisher is established outside the EU,
the EU authorized representative is:
**Springer Nature Customer Service Center GmbH**
**Europaplatz 3, 69115 Heidelberg, Germany**

Printed by Libri Plureos GmbH
in Hamburg, Germany